Electromagnetic Processing of Materials (EPM 2021)

IOP Conference Series: Materials Science and Engineering
Volume 1223

Online
14 – 17 June 2021

Editors:

Imants Kaldre
Arturs Brekis

ISBN: 978-1-7138-4681-9
ISSN: 1757-8981

Printed from e-media with permission by:

Curran Associates, Inc.
57 Morehouse Lane
Red Hook, NY 12571

Some format issues inherent in the e-media version may also appear in this print version.

Institute of Physics
Dirac House, Temple Back
Bristol BS1 6BE UK

Phone: 44 1 17 929 7481
Fax: 44 1 17 920 0979

techtracking@iop.org

Additional copies of this publication are available from:

Curran Associates, Inc.
57 Morehouse Lane
Red Hook, NY 12571 USA
Phone: 845-758-0400
Fax: 845-758-2633
Email: curran@proceedings.com
Web: www.proceedings.com

TABLE OF CONTENTS

IOP Conf. Series: Materials Science and Engineering 1223 (2022) 011001 doi:10.1088/1757-899X/1223/1/011001

Preface to IOP conference series issue
Electromagnetic Processing of Materials (EPM2021), June 14-17 (Online)

Special issue editors: Imants Kaldre, Artūrs Brēķis

Electromagnetic Processing of Materials (EPM) is a branch of research about the possibilities to improve various physical materials science processes with electric or magnetic fields. This covers areas such as liquid metal processing, metal casting and solidification under fields, induction heating and hardening. EPM also includes crystal growth from the melt, plasma processes, microwave processing recycling, separation. The series of EPM conferences was initiated in 1994 by S. Asai (Japan) and M. Garnier (France). Conference takes place every three years alternating between Europa and Asia (Nagoya 1994, Paris 1997, Nagoya 2000, Lyon 2003, Sendai 2006, Dresden 2009, Beijing 2012, Cannes 2015, Hyogo 2018, Riga 2021). Next conference is scheduled to take place in 2024 in Shanghai. The conference is open for any kind of materials processing involving electric or magnetic fields. It intends to bring together people from academic institutions, industry, and related equipment manufacturers.

Conference dates: June 13-17, 2021

Conference venue: Riga, Latvia, University of Latvia, Jelgavas street 3, LV-1004

Due to the travel and public event restrictions EPM2021 was held online in Zoom platform. We decided to keep the announced dates and make online event rather than postpone event to unknown time. It would be difficult to find good time slot for this conference between other postponed related conferences.

We attach the conference program. Time limits for each presenter were 30 minutes for Keynote and 20 minutes for regular report. Short discussion after each speech.

Participants from more than 10 countries. More than 50 scientific reports, more than 100 unique participants listening. Each session was chaired by two recognized scientists. We experienced very little technical difficulties during the conference sessions.

Conference Organizing team:

Imants Kaldre	Head of the organization committee	imants.kaldre@lu.lv
Leonīds Buligins	Chairman of the conference	leonids.buligins@lu.lv
Andris Jakovičs	Chairman of the scientific advisory board	andris.jakovics@lu.lv

IOP Conf. Series: Materials Science and Engineering 1223 (2022) 011001 doi:10.1088/1757-899X/1223/1/011001

Kristīne Kuzmane	Conference secretary	kristine.kuzmane@lu.lv
Svetlana Ščaņicina	Abstract and paper submission	svetlana.scanicina@lu.lv
Artūrs Brēķis	Technical support	arturs.brekis@lu.lv

Information: epm2021@lu.lv
Website: https://www.epm2021.lu.lv/

Imants Kaldre
EPM main Organizer
03.01.2022

IOP Conf. Series: Materials Science and Engineering 1223 (2022) 011002 doi:10.1088/1757-899X/1223/1/011002

Peer review declaration

All papers published in this volume of Electromagnetic Processing of Materials (EPM 2021) conference special issue have been peer reviewed through processes administered by the Editors. Reviews were conducted by expert referees to the professional and scientific standards expected of a proceedings journal published by IOP Publishing.

- **Type of peer review: Single-blind / ~~Double-blind / Triple-blind / Open / Other~~ (please describe) All articles were sent to independent reviewer, which is the expert in the field. After review, the reviewers form was sent back to authors. Corrected article was sent to reviewer again to check if corrections are implemented. Review processes were anonymous and all communication between reviewers and authors were carried out by conference e-mail. Reviewer form were used (see attached).**

- **Conference submission management system: https://www.epm2021.lu.lv/ (Review invitations and communication with authors were carried out via e-mail)**

- **Number of submissions received: 7**

- **Number of submissions sent for review: 7**

- **Number of submissions accepted: 6**

- **Acceptance Rate (Number of Submissions Accepted / Number of Submissions Received X 100): 86**

- **Average number of reviews per paper: 1.5 (Half of the accepted papers were accepted after one review round and minor revisions, half of the papers were sent to the reviewer two times, because reviewer still have some suggestions after corrections suggested in the first review)**

- **Total number of reviewers involved: 9**

- **Any additional info on review process: No**

- **Contact person for queries: Imants.kaldre@lu.lv**

3D numerical simulation with experimental validation of a traveling magnetic field stirring generated by a Bitter coil for silicon directional solidification process

B Hiba[1*], Ab Nouri[1], L Hachani[1] and K Zaidat[2]

[1]Laboratoire Physique de Matériaux, Université Amar Telidji de Laghouat,
BP 37 G Boulevard des martyrs, Laghouat 03000, Algérie.
[2]SIMAP-EPM PHELMA, University of Grenoble Alpes,
BP 75, 38402 Saint Martin d'Hères Cedex, France.

Email: b.hiba@ens-lagh.dz

Abstract. Silicon is the most widely used raw material in photovoltaic industry; however, the quality of the silicon photovoltaic solar cells depends on the quality of the raw material (i.e. metallurgical silicon or poly-silicon) and the solidification methods used for the production of the silicon ingot from which the solar cells are produced. This study is related to how improve the quality of the final ingot in the directional solidification process; it is necessary to control the impurity segregation of silicon raw material during the processing. This control can be accomplished by adding an electromagnetic Bitter coil which can generate an external traveling magnetic field (TMF) stirring to control the hydrodynamic flow of silicon melt during the solidification process without contaminating it. To carry out this study, we used a Bridgman vertical directional solidification furnace, equipped with a cylindrical Bitter coil stirrer used in order to have the control of the silicon melt convection on the principal parameters of the solidification process, such as the growth rate, the thermal gradient and the natural convection of silicon melt. For the electromagnetic, heat exchange and silicon melt flow modelling, we used 3D numerical Multiphysics coupled models. Parallel to the numerical results we carried out experimental investigations relating to the characterization of the electromagnetic parameters. This study shows a promising effect of the applied traveling magnetic field on the final ingot quality; indeed, we have the ability to control the silicon melt flow which can affect the thermal configuration, the solidification interface shape and the segregation of impurities by changing the electric current input configuration of the Bitter coil stirrer.

Key words: Bitter coil, silicon melt stirring, travelling magnetic field, vertical directional solidification.

1. Introduction

Many kinds of silicon solar cells are available in the market and the best in quality is made from high raw material quality like poly-silicon. This raw material increases the cost of the solar cells due to the fact that the raw material is the most expensive constituent. Our goal is to find a way to reduce this cost by using a cheap raw material, such as the upgraded metallurgical (UMG) silicon as solar-grade silicon (SoG) [1]. This silicon is less expensive than polysilicon due to its quality according to the impurities level; but it can be used for solar cells production with proper production process [2]. In addition, the growth rate, thermal gradient and fluid convection (natural or forced) are predominant factors that may affect the segregation of impurities in the final solidification process, whether for alloys [3-5] or pure material production. Based on this, it is necessary to control these factors in order to optimize the solar cells properties according to the purity level of the used raw material. The main important key investigated in this study is to control the segregation phenomena by controlling the silicon melt hydrodynamic flow and optimizing the shape of the silicon solid/liquid interface. Our approach is based on a numerical simulation coupled model supported by an experimental characterization. The study focuses on the different possibilities of controlling the hydrodynamic of the silicon melt using the interaction between the thermal configuration, imposed by the furnace, and the travelling magnetic field, imposed by the Bitter coil stirrer.

2. Materials and methods

To reach our goal, we used in this study a vertical Bridgman (2 Inches) furnace, denoted VB2, which operates in the resistive mode. VB2 is a cylindrical solidification furnace with an axisymmetric vertical axis [6-7]. In order to study deeply the directional solidification process of metallurgical photovoltaic silicon under the solidification key parameters, such as: thermal gradient and convection, VB2 furnace was equipped with specific thermal and hydrodynamic controls to modify the thermal gradient and the silicon melt convection. The thermal control is carried out by an instrumented heating system composed by two separate resistors; in the upper part of the furnace for the hot zone and in the lower part of the furnace for the cold zone. By this design, we have a separate thermal control over two zones that can generate a precise thermal gradient in the central zone of the furnace. The central zone, between the two resistors, was thermally insulated in order to create an adiabatic zone for the silicon melt charge according to the principle of Bridgman solidification process. For the hydrodynamic part, the control was obtained by the incorporation of an electromagnetic Bitter coil stirrer in the central zone (figure 1) which can generate a TMF configuration. This Bitter coil is composed of six blocks of ten disks to build the coils (figure 1 (b) and (c)); powered by a three-phase electric currents at frequency f_0 = 50 Hz and with a pulsation ω equal to $2\pi/T$ ($T = 1/f_0$ represent the period). All the components of the Bitter coil are installed in a stainless steel enclosure (figure 1 (a)) with the following dimensions: the internal radius is equal to 10 cm and the height is equal to 25 cm.

(a)	(b)	(c)

Figure 1. (**a**) Bitter coil stirrer (**b**) and (**c**) pictures of the six blocks that make up the coil.

The model was carried out by COMSOL Multiphysics software based on the presented design in figure 1; the schematic of 3D numerical model of the Bitter coil stirrer is shown in figure 2. This coupled model offers the possibility of analysing numerically the configurations of the electromagnetic field and the hydrodynamic flow of the silicon melt.

IOP Conf. Series: Materials Science and Engineering 1223 (2022) 012001 doi:10.1088/1757-899X/1223/1/012001

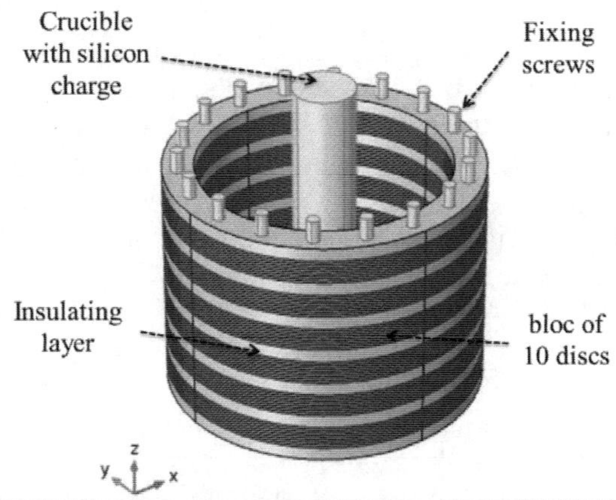

Figure 2. Schematic of the 3D numerical model of the Bitter coil stirrer.

As illustrated in figure 2, the silicon is contained in a fused silica cylindrical crucible and positioned in the central axis of the Bitter coil stirrer with the following dimensions: the radius $R = 2.5$ cm and the height $H = 30$ cm. The Bitter coil is composed of 6 blocks of 10 disks; each block is separated with an insulating layer, the whole system is assembled by well insulated fixing screws. For simulation purposes, the silicon charge was considered as liquid at a temperature above the silicon melting point ($T_m = 1687$ °K). The used model allowed us to perform different numerical configurations in order to discern the ability of the designed Bitter coil stirrer to generate an acceptable TMF configuration; which in turn, creates a better control of the silicon melt movement during the solidification process for the purpose to improve the segregation of the impurities in the final ingot. Our tests show that without the contamination of silicon melt, the TMF configurations can be changed externally by modifying the electromagnetic input parameters.

3. Traveling magnetic field (TMF)

The obtained traveling magnetic field is created by the three-phase alternating electric current with a phase shift of 120 degrees (2π/3) with adequate input currents. Base on this, the input current of the three phases (I_A, I_B and I_C) are expressed, in the equation system (1) figure 3, as function of the generator input currents (I_1, I_2 *and* I_3) and ω the pulsation [8].

Poly-phase input current configurations:

$$\begin{cases} I_A = I_1 \cdot cos(\omega t) \\ I_B = I_2 \cdot cos\left(\omega t + \dfrac{2\pi}{3}\right) \\ I_C = I_3 \cdot cos\left(\omega t + \dfrac{4\pi}{3}\right) \end{cases} \quad (1)$$

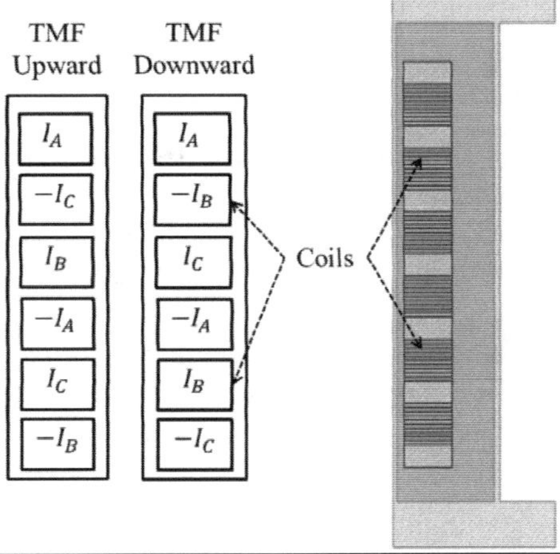

Figure 3. Illustration of the poly-phase input currents in the Bitter coils for upward and downward TMF configurations.

For the purpose of controlling the hydrodynamic flow of the silicon melt, we chose two specific configurations of TMF, the first one with a TMF upward and the second one with a TMF downward. These two configurations can be obtained by changing the distribution of the phases in the input current of the Bitter coil stirrer as illustrated in figure 3. Due to this three-phase system connection; the maximum value of the magnetic field density moves vertically in the middle of the Bitter coil stirrer, and therefore over the crucible and the melt silicon charge, as shown in figures 4 and 5 for the used conditions: $f_0 = 50$ Hz and the input currents of the three-phase are: $I_1 = I_2 = I_3 = 300$ A.

4. Results and discussions

4.1. Numerical simulation results

For the numerical results, we started with the presentation of the electromagnetic filed configuration. In figure 4, we can see the numerical results of the 3D field configuration of the magnetic flux intensity.

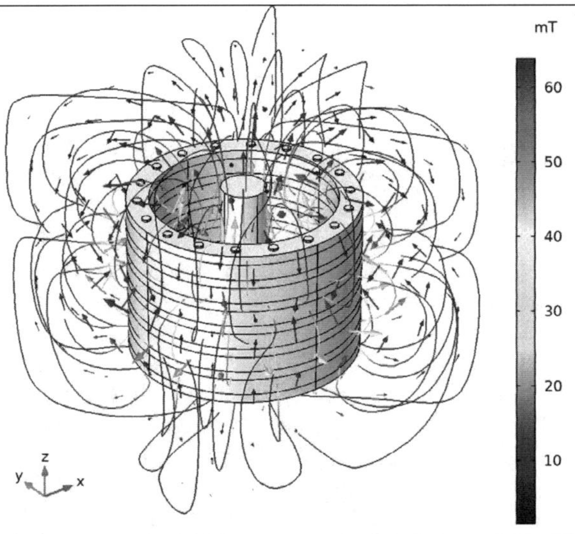

Figure 4. 3D simulation results of the configuration of the magnetic flux streamline of the TMF downward case for the following operating conditions: $f_0 = 50$ Hz and the input currents are: $I_1 = I_2 = I_3 = 300$ A.

Since the Bitter coils are placed along the *OZ* axis; the created magnetic field is directed along this axis; the streamlines of the magnetic flux intensity show the existence of two loops along the Bitter stirrer (see figure 5). If the coils of the Bitter stirrer are correctly supplied by a poly-phase electric source with a pulsation ωt (as presented in equation (1) and illustrated in figure 3), the magnetic field propagation should be in form of a traveling wave parallel to the *OZ* axis. To understand this result, figure 5 shows the geometry of the central section of the system by the illustration of the vertical cutting median plane through the Bitter coils and the silicon melt charge in the centre. These results are for the downward case, they show the TMF configuration with the norm of the magnetic flux intensity at the following operating conditions for the Bitter coil: the frequency $f_0 = 50$ Hz and the input currents for the three-phase are: $I_1 = I_2 = I_3 = 300$ A. The difference between the three cases illustrated side by side pictures in figure 5 is related to three different spatial presentation of the magnetic field; suggesting that the designed Bitter coil stirrer has the ability to generate an acceptable movement of the TMF wave that can be used for the stirring purpose of the silicon melt charge.

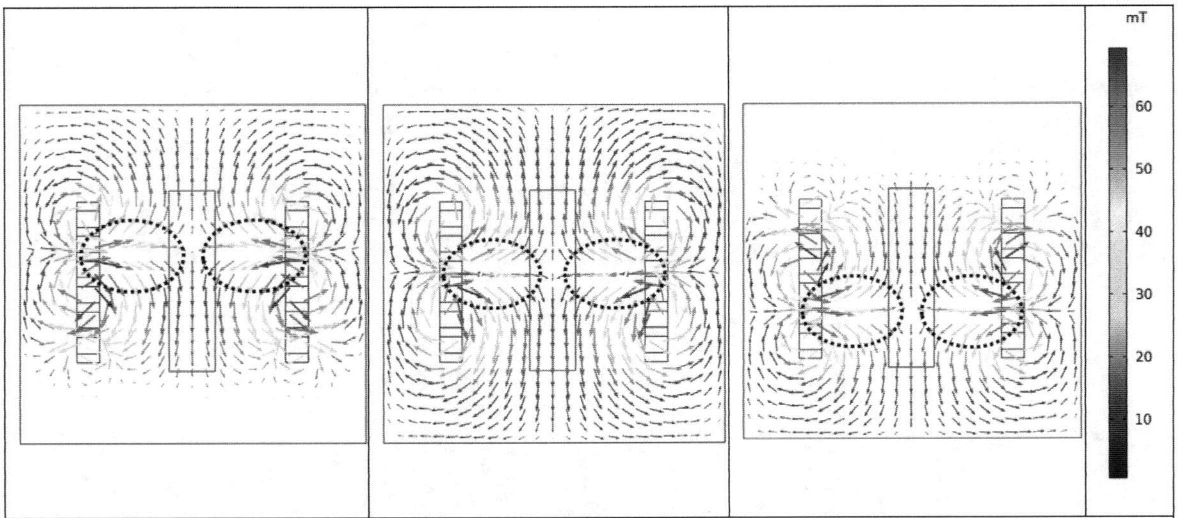

Figure 5. Numerical results of the spatial evolution of arrow vectors field of the magnetic flux intensity norm for the downward TMF case for the following operating conditions: f_0 = 50 Hz and the input currents are: I_1= I_2= I_3= 300 A.

The three pictures in figure 5 show the numerical results of the arrows field of the magnetic flux intensity norm. Those pictures, from left to right, illustrate three consecutive instant in the same period (*T*). The maximum of the magnetic flux intensity is located near the coils and moves from the top to the bottom of the Bitter coil as shown by the dotted circles in black colour (TMF downward case). This movement is related to the phase shift in the input parameters of the three-phase system between the three instants. To present this phase shift configuration, in figure 6 we plotted the dimensionless current from the equation system (1) as a function of time for the three-phase.

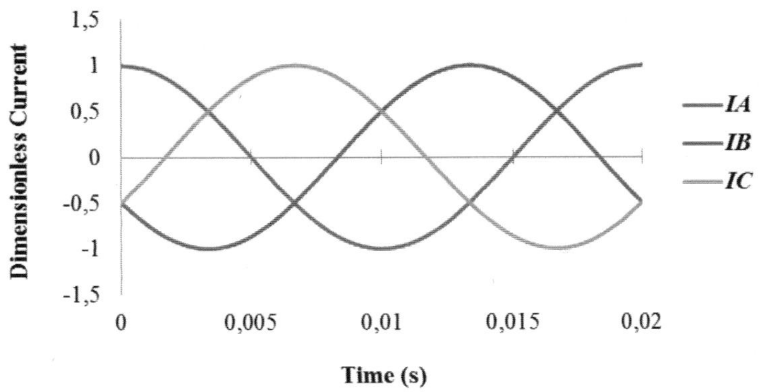

Figure 6. Dimensionless configuration of the three-phase current as a function of time according to the system phase shift for one period $T = 1/f_0$.

As we can see, according to the system equation (1), figure 6 shows the phase shift and the variation of the electric input current as a function of time between the there-phase, the result of this phase shift configuration with the connections system used in figure 3 allow the creation of the adequate TMF movement.

This Bitter coil conception offers the ability to control the value of the magnetic field intensity in the silicon melt by varying the values of the input currents; moreover, it is also possible to change the direction of the TMF effect by modifying the shift of the positions of the three phases of the Bitter coils. Likewise, for the upward case, the TMF wave has the same movement but from the bottom to the top of the Bitter coil (upward). Therefore, the control of the hydrodynamic flow in the silicon melt is possible by magneto hydrodynamic coupled interaction for both cases: a TMF and hydrodynamic flow upward and a TMF and hydrodynamic flow downward in the vicinity of the crucible wall. The 3D numerical results of the magneto hydrodynamic flows of molten silicon for the two TMF configurations are presented in figure 7:

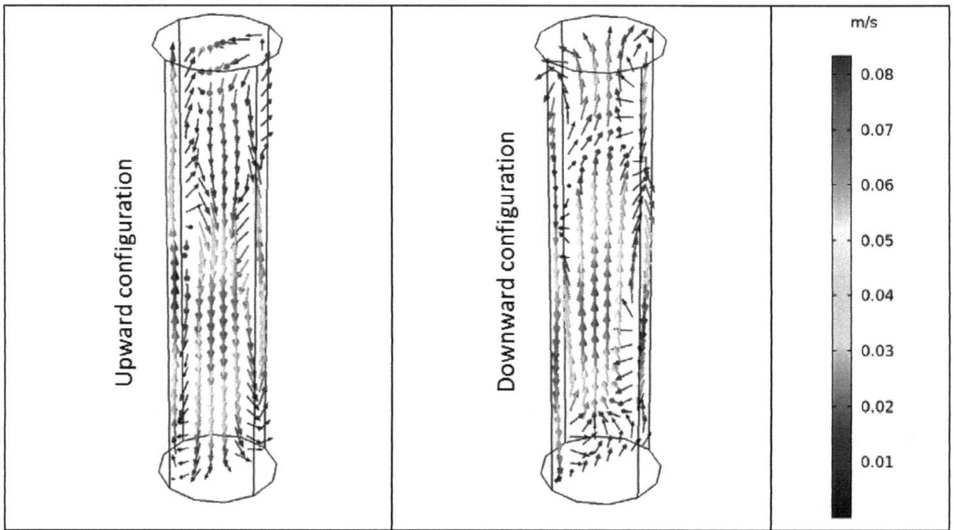

Figure 7. 3D numerical results of velocity vector fields of the hydrodynamic flow of silicon melt for TMF configurations for the operating conditions: $f_0 = 50$ Hz and $I_1 = I_2 = I_3 = 300$ A.

As shown in figure 7, we can see the direction of the silicon melt convection and the velocity obtained for two hydrodynamic flow configurations: the result on the right illustrates the downward hydrodynamic flow for the TMF downward configuration and the result on the left illustrates the upward hydrodynamic flow for the TMF upward configuration. For both cases and for the same input parameters of currents we obtained the same maximum value of the flow velocity with the installation of two pairs of vortices. Consequently, depending to the initial imposed thermal configuration and natural hydrodynamic flow; the Bitter coil stirrer effect can provide better control over the key parameters of the impurities segregation phenomena, such as: the velocity of the silicon melt and the shape of the solid-liquid interface during the solidification process. By having sufficient flow velocity of silicon melt and an almost plan and horizontal solid/liquid interface shape, we can significantly improve the segregation of impurities in the final ingot [6-7].

4.2. Experimental validation

In this section, we present the validation of the electromagnetic parameters of our Bitter coil stirrer. The numerical results of the magnetic intensity are compared to the electromagnetic experimental measurements (figure 8). The Bitter coil stirrer is powered by a three-phase AC with the frequency $f_0 =$ 50 Hz.

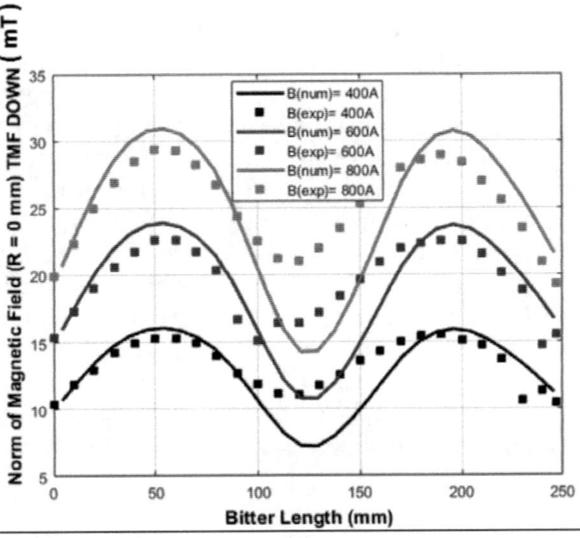

(a)

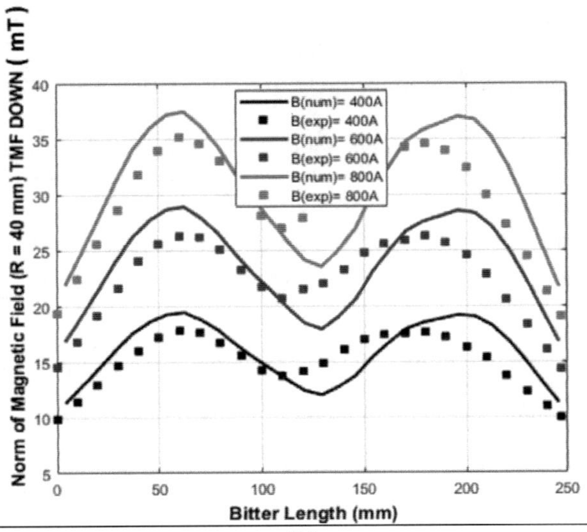

(b)

Figure 8. Comparison between numerical results and experimental measurements of the magnetic intensity on the vertical direction *OZ* at the centre position of the Bitter coil stirrer for: $R = 0$ cm and $R = 4$ cm with $f_0 = 50$ Hz and $I_1 = I_2 = I_3 = 400$; 600 and 800 A.

Figure 8 shows the results of the electromagnetic intensity in z-positions (i.e. in vertical direction) for the two following positions: (a) for the z position equivalent to the radius $R = 0$ mm and (b) for the z position equivalent to the radius $R = 40$ mm. To ensure that the same results are obtained regardless the input current, measurements were carried out for different applied values: $I_1 = I_2 = I_3 = 400$; 600 and 800 A. The electromagnetic results obtained by the model are in concordance with the experimental results; consequently, they can be used for the validation of a future experimental investigation on the generated hydrodynamic flow in the silicon melt.

IOP Conf. Series: Materials Science and Engineering 1223 (2022) 012001 doi:10.1088/1757-899X/1223/1/012001

5. Conclusions

The electromagnetic configuration and the hydrodynamic flow in the silicon melt were numerically and experimentally investigated. The obtained results show clearly the possibility to control the silicon melt convection, by adding an external traveling magnetic field stirring apparatus, to control the silicon melt convection without physical contact through the magneto hydrodynamic volume forces (i.e. Lorentz forces). Good concordance between the obtained numerical results and experimental results validated our numerical model that can be used as an efficient tool to investigate and predict the possible forced configuration using our traveling magnetic field Bitter coil stirrer. The Bitter coil stirrer has been designed to generate different modes of traveling magnetic field (i.e. upward or downward stirring); each configuration giving a specific electromagnetic force pattern with a particular hydrodynamic flow affecting directly the thermal configuration and the shape of the solid/liquid interface. It was shown that it is possible to change the melt flow convection by using either upward or downward stirring configuration; such result will have then a direct effect on the impurity segregation as well. Further investigation will be useful to quantify the effect of such melt flow modification on the segregation and the distribution of impurities on the scale of the whole silicon ingot.

6. References

[1] Meyer S, Wahl S, Molchanov A, Neckermann K, Moller C, Lauer K and Hagendorf C 2014 *Solar Energy Materials & Solar Cells* **130** 668.

[2] Fornies E, Del-Canizo C, Mendez L, Souto A, Vazquez A P and Garrain D 2021 *Solar Energy* **220** 354.

[3] Noeppel A, Ciobanas A, Wang X D, Zaidat K, Mangelink N, Budenkova O, Weiss A, Zimmermann G and Fautrelle Y 2010 *The Minerals Metallurgical and Materials Transactions* **B 41** 193.

[4] Hachani L, Saadi B, Wang X D, Nouri Ab, Zaidat K, Belgacem-Bouzida A, Ayouni-Derouiche L, Raimondi G and Fautrelle Y 2012 *International Journal of Heat & Mass Transfer* **55** 1986.

[5] Hachani L, Zaidat K and Fautrelle Y 2015 *International Journal of Heat & Mass Transfer* **85** 438.

[6] Cablea M, Zaidat K, Gagnoud A, Nouri Ab and Delannoy Y 2014 *Journal of Crystal Growth* **401** 883.

[7] Cablea M, Zaidat K, Gagnoud A, Nouri Ab, Chichignoud G and Delannoy Y 2015 *Journal of Crystal Growth* **417** 44.

[8] Hachani L 2013 Etude de l'influence de la convection naturelle et forcée sur le processus de la solidification : cas d'un alliage métallique binaire Thesis Grenoble-INP university France.

Electromagnetic Processing of Materials (EPM 2021) IOP Publishing
IOP Conf. Series: Materials Science and Engineering 1223 (2022) 012002 doi:10.1088/1757-899X/1223/1/012002

Electromagnetically coupled processes – An error estimator for electromagnetic computational models

Jesus Garcia[1,2], **José Alves**[2], **Julien Barlier**[2], **Grégory Puaux**[2], **François Bay**[1]

[1]Cemef, Mines ParisTech PSL Research University, CNRS UMR7635,
CS 10207, Rue Claude Daunesse, Sophia Antipolis cedex 06904, France

[2]Transvalor S.A. E-Golf Park, 950 Avenue Roumanille, 06410 Biot, France

E-mail: francois.bay@mines-paristech.fr

Abstract. Reliable modelling of electromagnetic processing technologies largely depends on the accuracy of electromagnetic computations. This work will thus focus on the development of error estimators for evaluating the accuracy of finite element electromagnetic computations dedicated to improving induction heating process modelling accuracy. Application to induction heating and magnetic stirring cases will prove the efficiency of the approach. Error estimation results will be used later on for carrying out anisotropic adaptive remeshing.

1. Introduction

Electromagnetically coupled processes are increasingly used in the manufacturing industry. Induction heating processes for instance are commonly used for initial bulk heating or surface heat treatment. The computational modelling tools for designing these processes require dealing with electromagnetic/heat transfer couplings, as well as metallurgical evolution and mechanical distortions for residual stress prediction. However, one of the problems with the finite element tools used for modelling these processes is to make sure that these tools can be used in a reliable way. Several estimators have been proposed in literature for eddy current problems, but most of them deal with harmonic formulations ([1]). We aim here at introducing an error estimator suited for time-dependent formulations of the electromagnetic problem in fully immersed FE approach.

As electromagnetically coupled processes often involve non-linear magnetic effects, we have developed a model ([2]) which takes into account these effects using a full-time integration scheme. We shall introduce here an error estimator suited to our model. We have then tested it in an induction heating case as well as a magnetic stirring case, and carry out an error analysis by studying convergence as a function of varying mesh size.

IOP Conf. Series: Materials Science and Engineering 1223 (2022) 012002 doi:10.1088/1757-899X/1223/1/012002

2. The computational model

The computational model is based on a coupling between the electromagnetic model and addition al physical models – such as heat transfer, solid mechanics, ….

2.1. The electromagnetic model

The electromagnetic model is classically based on the Maxwell equations (1) completed by the electromagnetic constitutive laws (2).

$$\begin{cases} \vec{\nabla} \times \vec{H} = \vec{J} \\ \vec{\nabla} \times \vec{E} = -\dfrac{\partial \vec{B}}{\partial t} \\ \vec{\nabla} . \vec{B} = 0 \end{cases} \tag{1}$$

$$\begin{cases} \vec{B} = \mu(\| \vec{H} \|, T)\, \vec{H} \\ \vec{J} = \sigma(T)\vec{E} \end{cases} \tag{2}$$

2.2. The numerical approximation for the electromagnetic problem

The previous equation system is solved using an (A,Φ) formulation.

$$\begin{cases} \sigma \dfrac{\partial \vec{A}}{\partial t} + \vec{\nabla}\left(\dfrac{1}{\mu}\, \vec{\nabla} \times \vec{A}\right) = -\sigma \vec{\nabla} \Phi \\ \vec{\nabla}(\sigma(\dfrac{\partial \vec{A}}{\partial t} + \vec{\nabla}\Phi))=0 \end{cases} \tag{3}$$

Space discretization is based on a global finite element discretization approach where the coil, the workpiece as well as the air are meshed with P1 edge-based finite elements ([3]).
This means that the magnetic potential vector field A is approximated as A^{P1} by constant values on each edge of the mesh – while the electric potential is approximated as Φ^{P1} by its values at the mesh nodes.

2.3. Time discretization and coupling with other physical problems involved in EPM processes

For time discretization, we have chosen to use a full-time integration scheme for the electromagnetic problem instead of a harmonic scheme – in order to deal with non-linear magnetic materials. In a coupled Multiphysics problem as this one, we typically have two specific time scales: the first one for the electromagnetic problem - typically a fraction of the current period 10^{-2} down to 10^{-6} s. -– while the second is for the thermomechanical problem - typically 1 s. Coupling with the heat transfer equation is carried out here by estimating an average dissipated Joule heat source term and using it over several thermomechanical time steps.
Coupling with the other physical problems involved is carried out by computing the average Joule heat source term due to eddy currents and of instantaneous Lorentz forces (4)

$$\begin{cases} P_{eddy} = \dfrac{1}{T}\displaystyle\int_0^T \sigma \left\| \dfrac{\partial \vec{A}}{\partial t} \right\|^2 \\ \vec{F}_{Lorentz} = \sigma \dfrac{\partial \vec{A}}{\partial t} \times (\vec{\nabla} \times \vec{A}) \end{cases} \tag{4}$$

3. Error estimators for the electromagnetic problem

Error estimation for electromagnetically coupled processes arises from the multi-physical coupled problems. It exists various error sources. It could be solving the electromagnetic problem, from the heat transfer model, in case of metallurgical couplings, error from the metallurgy model, in case of mechanical couplings, error from the solid mechanics model.
This paper is focused on the error that occurs when solving only the electromagnetic model.
There are two main kinds of error estimators: a priori estimators – based on extrapolation of solutions obtained with mesh refinement or interpolation increase - or a posteriori estimators – based on postprocessing of numerical solutions. We shall focus here on this second category.
There are three main kinds of a posteriori estimators: residual-based estimators, equilibrated estimators and recovery-based estimators. We present here an estimator based on this last approach.

The main idea behind recovery-based estimators is to build a higher interpolation order field compared to the one calculated with the finite element approximation. This approach for a posteriori estimators was introduced by Ainsworth and Craig[4].
In the electromagnetic model, we obtain an elementwise approximated solution of the magnetic potential vector (A^{P1}) in order to obtain a more accurate representation of its first-order curl-derivative – namely H^{P0}.

The strategy we have thus developed is based on the following stages:

- Computation of the A^{P1} solution
- Derivation of the H^{P0} magnetic field: curl(A^{P1})
- Recovery of an H^{P1} magnetic field
- Computation of the (H^{P1} - H^{P0}) difference on the mesh

Two strategies can be carried out for carrying out the recovery operation: a first one based on Super Convergent Patch Recovery (SPR method) and another based on Recovery by Galerkin-based method. This last method proves to be more robust in terms of numerical convergence and efficient in terms of CPU time. Hence, the approach for the computation of the recovered solution in this work is based on a Galerkin minimization method.

The H^{P0} magnetic field is derived using the curl of the A^{P1} vector magnetic potential field; the values for the H^{P0} field are computed and then stored at the Gauss integration points. Then, using the recovery method the H^{P1} field is computed on the edges starting from the H^{P0} field and the edge element shape functions. This minimization stage leads to solving a finite element system with a mass matrix and a load vector, as follows

$$\langle \psi_j, \psi_i \rangle \{H_e^{P1}\} = \langle \psi_i, H^{P0}(X_{GaussPts}) \rangle \tag{5}$$

To compute the H^{P1} field values on the integration points, an interpolation of the H^{P1} field on the edges is carry out, as follows

$$H^{P1}(X_{GaussPts}) = \sum_{e=1}^{nb_{edges}} H_e^{P1} \psi_e(X_{GaussPts}) \tag{6}$$

Finally, the estimator is computed as follows

$$\epsilon = \| H^{P1} - H^{P0} \| \tag{7}$$

4. Applications

This methodology has been applied to several cases involving electromagnetic couplings. These cases have been carried out within the framework of the commercial software FORGE® & THERCAST®. The estimators show a good convergence rate with the mesh size. We present here an application to an induction heating case modelled with FORGE®, as well as a magnetic stirring case modelled with THERCAST®.

4.1. An induction heating case

The case investigated here is the TEAM 3 (TEAM (Testing Electromagnetic Analysis Methods) represents an open international working group aiming to compare electromagnetic analysis computer codes) problem; it consists of a conducting ladder, having two holes with a coil above carrying a sinusoidal current.

Figure 1-a. The case

Figure 1-b. The associated mesh

Figure 1-a displays the geometry of both the part to be heated as well as the inductor. Meshes for the part and inductor are displayed in Figure 1-b. It should be noted that these meshes are embedded in a global mesh modelling the surrounding air.

Figure 2 displays the magnetic field distribution after FE calculation on the part.

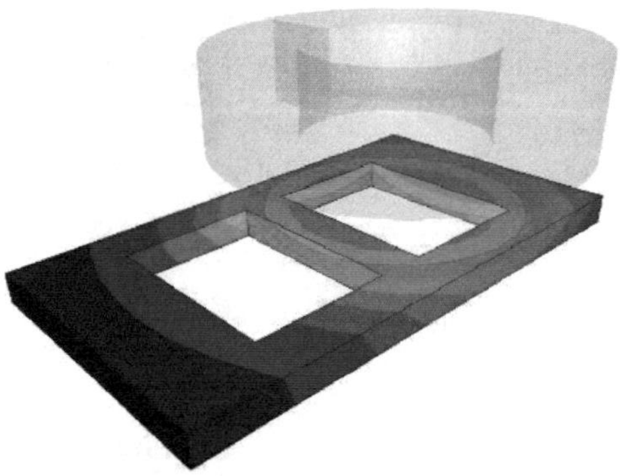

Figure 2. Magnetic field distribution on the part to be heated.

The next stage is to use the finite element error estimator introduced previously in order to get a first estimate of error values. Moreover, we know from finite element theory that numerical errors should decrease when the average mesh size is reduced. We have thus carried out a mesh sensitivity analysis in order to check the convergence of our estimator. Meshes based on a decreasing mesh size are displayed in Figure 3a.

Figure 3b presents the errors distributions obtained with the estimator. As one can see, larger errors can be observed in the coarser mesh, as well as a continuous decrease as the mesh gets finer. It can thus be clearly seen that the finite element computational error vanishes with decreasing mesh size, which is coherent with finite element theory.

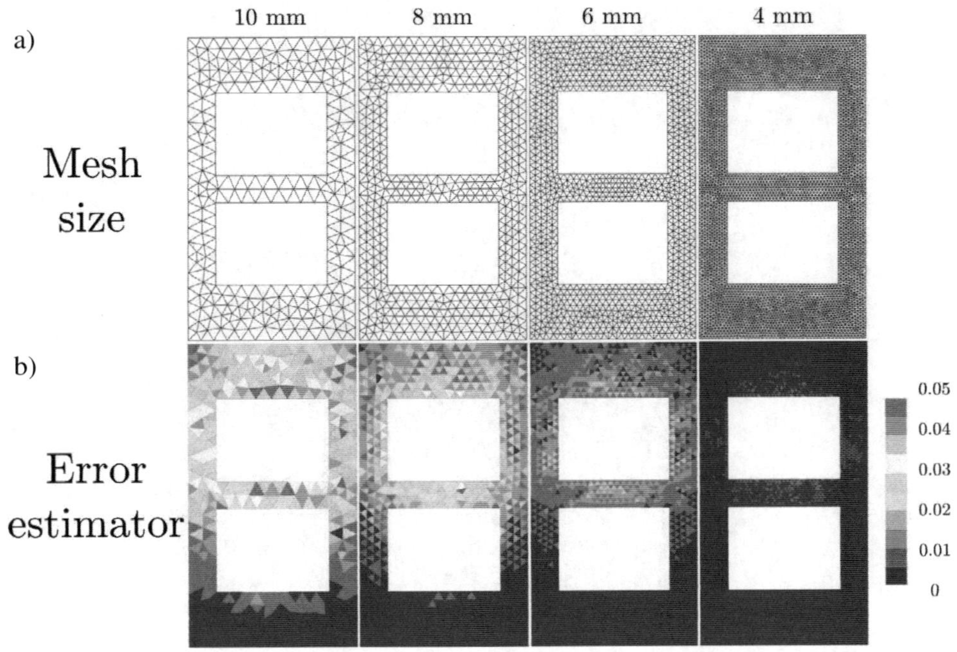

Figure 3. a (Top) Mesh size refinement; b (Bottom) Associated error

4.2. A magnetic stirring case

The benchmark used to test the estimator on the electromagnetic stirring process is based on the results proposed by Musaeva [5]. The case consists in a laboratory scale stirring application of Galinstan melt in a Plexiglas mould.

Figure 4 displays the geometry, as well as the associated mesh; thanks to the axial symmetry of the geometries, a reduced 15° angular section has been simulated.

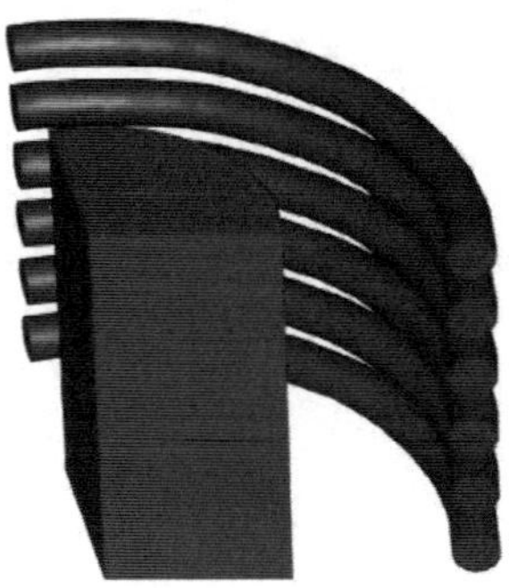

Figure 4-a. The case

Figure 4-b. The associated mesh

IOP Conf. Series: Materials Science and Engineering 1223 (2022) 012002 doi:10.1088/1757-899X/1223/1/012002

As for the previous case, a sensitivity analysis has been carried out to analyze the behavior of our error estimator as the mesh gets refined.Therefore, we have carried out computations on an increasingly refined sequence of meshes.

Figure 5a shows the increasingly refined meshes, while Figure 5b displays the errors obtained at a selected time step with the estimator. Here again, the errors keep decreasing as the mesh gets refined.

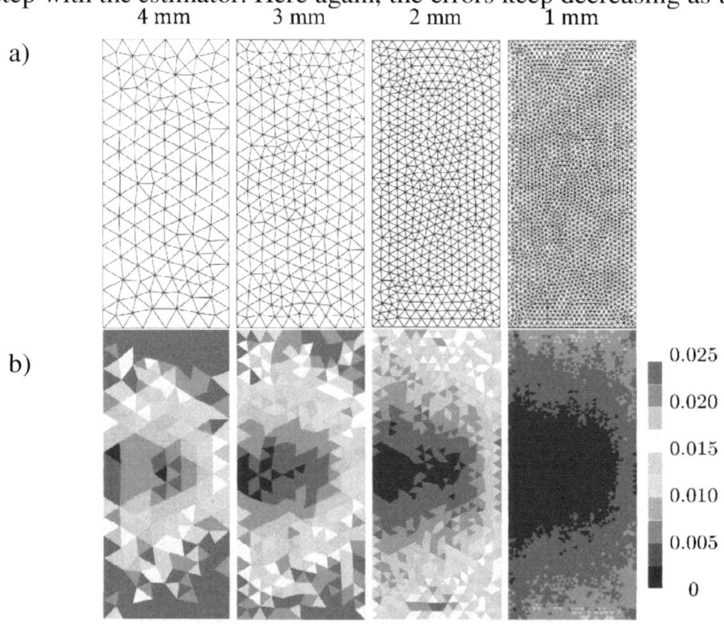

Figure 5. a (Top) Mesh size refinement; b (Bottom) Associated error

Figure 6 provides additional information by providing the error distribution for a selected mesh (mesh size – 2 mm) at different time steps of the simulation (the selected time steps are plotted on the sine wave curve).

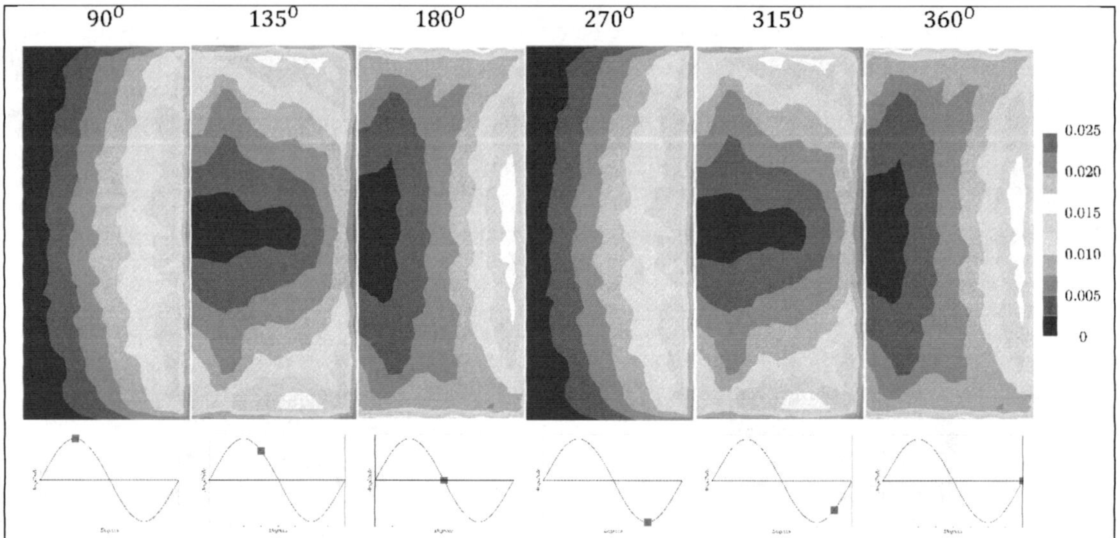

Figure 6. Different time steps following the sine wave and associated error (mesh size = 2mm)

The results show how the error distribution changes over space and time in conjunction with the travelling magnetic force field.

5. Conclusion

We have introduced in this paper a recovery-based a posteriori estimator. This estimator has been tested on the solutions provided by our computational model – which is based on a fully immersed finite element method approach in conjunction with a full-time integration of the Maxwell equations. The recovered solution is computed using a Galerkin-based approach.

This estimator has been tested on two electromagnetically coupled processes – an induction heating case, as well as a magnetic stirring case. The results on the induction heating case show a satisfactory behaviour; it enables to localise in space the areas with the larger errors and shows how these errors converge as the mesh gets refined. Moreover, the results for the magnetic stirring case show how the estimator manages to dynamically follow the magnetic field at every time step and thus enable the detection of the areas most affected by the error.

This error estimator will now be used in conjunction with an automatic adaptive remeshing algorithm in order to perform computations for which the errors comply with a specific tolerance defined by the end-user.

6. References

[1] Creusé E., Le Menach Y., Nicaise S., Piriou F., and Tittarelli R., "Two guaranteed equilibrated error estimators for Harmonic formulations in eddy current problems," *Comput. Math. with Appl.*, vol. 77, no. 6, pp. 1549–1562, 2019, doi: 10.1016/j.camwa.2018.08.046.

[2] Alves, J., and Bay, F. (2016). Modelling and Analysis of Electromagnetism in Magnetic Forming Processes. *IEEE Transactions on Magnetics*, Vol 52, N° 5

[3] Nédélec J.C., A new family of mixed finite elements in IR3. *Numer. Math.*, 50, 57-81.,1986

[4] M. Ainsworth and Craig A., "A posteriori error estimators in the finite element method," *Numer. Math.*, vol. 60, no. 1, pp. 429–463, Dec. 1991, doi: 10.1007/BF01385730.

[5] D Musaeva, V Ilin, E Baake, V Geza, Numerical simulation of the melt flow in an induction crucible furnace driven by a Lorentz force pulsed at low frequency, *Magnetohydrodynamics* 51 (4), 771-783

IOP Conf. Series: Materials Science and Engineering 1223 (2022) 012003 doi:10.1088/1757-899X/1223/1/012003

Model experiments for Czochralski crystal growth processes using inductive and resistive heating

A Enders-Seidlitz, J Pal, and K Dadzis

Leibniz-Institut für Kristallzüchtung (IKZ), Berlin, Germany

E-mail: josef.pal@ikz-berlin.de

Abstract. The Czochralski (CZ) growth technique is widely applied in crystal growth, using both induction and resistance heaters. In this work, a novel model experiment platform with comprehensive in-situ measurement capability is introduced. Growth experiments with the model material tin applying both heating concepts are performed and analyzed, e.g., in terms of the maximum achievable crystal diameter. Strong asymmetries in the magnetic field of the induction heater are measured and temperature distribution on the resistance heater is found to be non-uniform. Furthermore, significant losses are observed in the power supplies of the resistance heater. The heating efficiency of both concepts is compared considering different insulation geometries. The obtained results show the capability of model experiments for design optimization and will provide valuable input for further validation of numerical simulations.

1. Introduction

Crystals are an indispensable part of modern electronics. One of the main methods for crystal growth from the melt is the Czochralski (CZ) technique. Generally, the CZ process starts with the melting of the raw material in a crucible and continues with contacting of a single crystal seed to the melt surface. Under rotation and simultaneous upward pulling, the melt solidifies at the seed and a crystal grows. Detailed descriptions can be found in [1,2] and the references therein. Nowadays, the CZ technique is widely established to produce many types of bulk single crystals, e.g. Si or Ga_2O_3.

Both induction heating [3,4] and graphite resistance heaters [5,6] are applied in the CZ growth technique, selected based on various requirements such as cost efficiency, electromagnetic forces, or degree of purity inside the furnace. Numerical simulations have been applied for investigation of both concepts [4,6]. However, due to various simplifications in the numerical models as well as due to often uncertain or unknown material properties, validation is required to give reliable simulation results [7].

Setups on laboratory scale with low working temperatures, relaxed vacuum-sealing requirements and hence with easy access for various in-situ measurement techniques allow to investigate selected physical phenomena extracted from the complex crystal growth process. Such "model experiments" are, therefore, predestined for the validation of numerical simulation and are of growing interest, e.g. for investigation of flow phenomena [8-10], process control [11,12], or phase boundary stability [13]. Within the NEMOCRYS (Next Generation Multiphysical Models for Crystal Growth Processes) project, we are developing a new generation of multiphysical models [14,15] and model experiments for crystal growth processes, where multiple physical phenomena can be observed simultaneously and hence validated in a coupled manner. The current experimental basis for model experiments is the Test-CZ setup – a multipurpose system with improved similarity to real CZ growth furnaces. For the present study, the Test-CZ setup can be optionally equipped with an induction or a graphite resistance heater. Tin (Sn) is applied as a model material, selected due to its low melting point of 232°C and positive results in previous studies, e.g. for investigation of diameter control [12].

2. Experimental setup and measurement techniques

The Test-CZ setup consists of a steel furnace equipped with a top and bottom axis, a surrounding support structure, and a control unit for various electronic systems (see figure 1, left). The furnace is made of stainless steel and has an inner diameter and height of 263 mm and 383 mm, respectively. A large front door allows convenient in-situ access while various flanges are available, e.g., for connection of various sensors or heater power supplies. Different atmospheric conditions such as air (with open / closed door), vacuum, inert N_2 / Ar gas, or air cooling from the side using a fan can be applied in the experiment. Both furnace and axes are water-cooled, enabling the application of model materials with higher melting points in future.

The furnace can be equipped with two different heating systems. The induction heater consists of a copper coil with three windings and an inner diameter of 140 mm, that is internally cooled with water. It is powered by a generator (Trumpf, TruHeat MF 5030) delivering up to 30 kW power at frequencies between 5 to 200 kHz. The meander-shaped resistance heater is made of graphite, has an inner diameter of 110 mm and a resistance of 125 mΩ at room temperature. It is supplied from a transformer and thyristor power controller with a maximum of 4.2 kW at a voltage of 22 V and a frequency of 50 Hz. The latter setup can be optionally equipped with an insulation made of a carbon felt that can be further varied in its shape (without, bottom-only, side / different heights, complete including a cap). Due to the continuous development of the experiment the graphite crucibles used in the two setups are different: an inner diameter / height of 100 mm / 40 mm is used in the induction heating, and of 50 mm / 20 mm in the resistance heating setup.

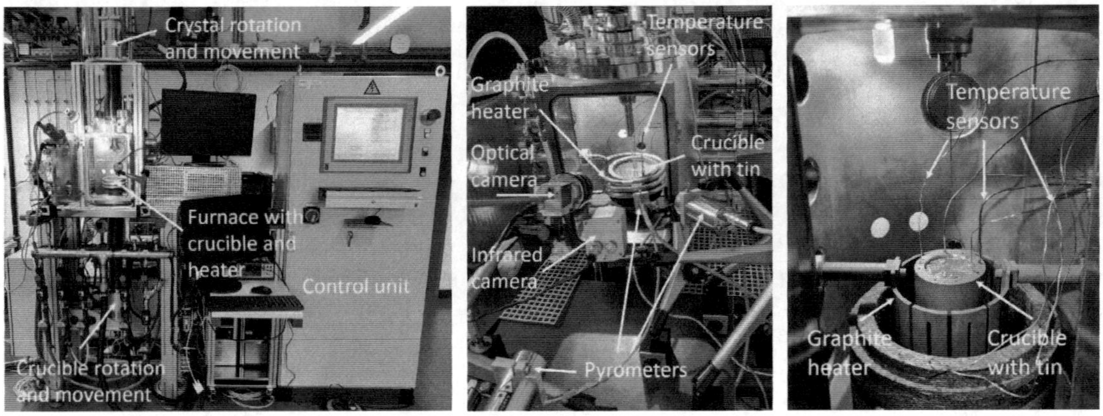

Figure 1. Picture of the Test-CZ laboratory setup: overview (left), cases with the induction heater (center) and the graphite heater (right). In the latter case, a partial insulation up to half of the total available height is used.

The Test-CZ setup is equipped with different measurement instruments and sensors for in-situ observation, ranging from optical images through temperatures to voltages, currents and magnetic fields. Some of the applied sensors are highlighted in figure 1 (center / right). In the following, the most important measurements for the present study are discussed in more detail.

Direct contact temperature sensors [16] of resistance type (4-wire Pt100) and sheath thermocouples (TC; type K, J) are mounted in the sidewall of the crucible (and insulation) to monitor its temperature distribution. Two temperature sensors are immersed directly into the melt with a distance of approx. 5 mm to the crucible bottom and wall: a Pt100 is applied for the temperature control using an external device (Eurotherm, model 902), while a TC at the same position serves for monitoring purposes. For

IOP Conf. Series: Materials Science and Engineering 1223 (2022) 012003 doi:10.1088/1757-899X/1223/1/012003

high-accuracy measurement, an external cold junction unit (Isotech, model TRU 937/50) is used in the connection of the TC's to the measuring instrument (Keithley, DAQ 6510 with 7710). Furthermore, all TC's and Pt's are calibrated in an external dry calibrator (Isotech, model Jupiter) in the range between 50°C and 400°C in steps of 50°C, which allows to achieve an accuracy of at least 0.2°C.

For contactless temperature measurement [16], various pyrometers are applied: short wavelength (2.0 - 2.6 µm, Lumasense, IGA 6-23 advanced), short wavelength ratio type (1.5 - 1.6 and 2.0 - 2.5 µm, Lumasense, IGAR 6 advanced), and a pyrometer array (Lumasense, series 600) equipped with 2 long wavelength (8 - 14 µm) receivers. The availability of different devices / wavelengths gives a greater scope for measurements of various materials in a broad range of temperatures; furthermore, the ratio pyrometer enables emissivity-independent temperature measurements under specific conditions such as insignificant wavelength dependency of the emissivity. In some cases, stickers with a known emissivity of 0.95 (8 - 14 µm) are applied. Additionally, an infrared (IR) camera [17] (Optris, PI-640) is used for recording of thermal images at 640x480 pixel resolution and wavelength of 8 - 14 µm. An optical camera (Basler acA2500-20gc, Kowa LM50HC lens) is used for documentation purposes.

For the analysis of the effective heater power, the simultaneous measurement of the heater current and voltage is required. In both setups, the current is measured using a Rogowski coil (Chauvin Arnoux MA200). In the resistance heating setup, measurement of the voltage drop on the graphite heater is sufficient; it is recorded using a multimeter. In contrast, in the induction heating setup, the phase shift between the current and voltage is required in addition to the voltage; respective measurements are under preparation. The magnetic field of the induction coil is measured using a Teslameter (Lakeshore, model F41) with a high-frequency probe.

Data acquisition is realized by a self-developed python program. For most of the measurement devices, such as the multimeter, the pyrometers, or the optical camera, separate interface libraries according to the respective connection (Serial RS-232 or RS-485 using an ethernet-to-serial converter Moxa Nport 5650-8-DT, Ethernet, USB) are implemented. A graphical user interface for process observation and visualization is realized using the PyQt5 library. The code is available under a GPLv3 open source license and can be found on GitHub [18].

3. In-situ analysis of the heaters

A series of experiments is performed to investigate the thermal and electromagnetic behavior of both heaters (cf. [3,5]), to get input data for validation of numerical simulations, as well as to analyze the advantages and disadvantages of both heating concepts applied in a similar setup.

A typical IR camera snapshot during a growth experiment with induction heating is shown in figure 2 (left). It can be seen that heat is directly generated in the crucible – where it is needed – while the coil itself remains cold. The melt temperature is controlled at 236°C by an applied current of $I = 162.3$ A at $f = 13.5$ kHz. Because of the missing voltage measurement, a 2D numerical simulation is used for power estimation [14,15]; it yields a heater power of 105 W. TC measurements in the crucible wall are indicated by arrows in figure 2 (left), showing a temperature asymmetry of 1.0°C.

The measured vertical component of the magnetic field along the inductor diameter is shown in figure 2 (right). To give a better comparison to a 2D simulation, the idealized symmetric distribution is indicated. Comparison of the two lines indicates a strong 3D character, as observed also in [4], thus the accuracy of a 2D simulation is limited. The asymmetric magnetic field is assumed to be caused by the influence of the current supplies at one side of the coil and by the spiral shape of the windings; further investigation using both additional measurements and 3D simulations are under preparation. A correlation between magnetic field and the temperature asymmetry is assumed, but cannot be assessed clearly with the available data.

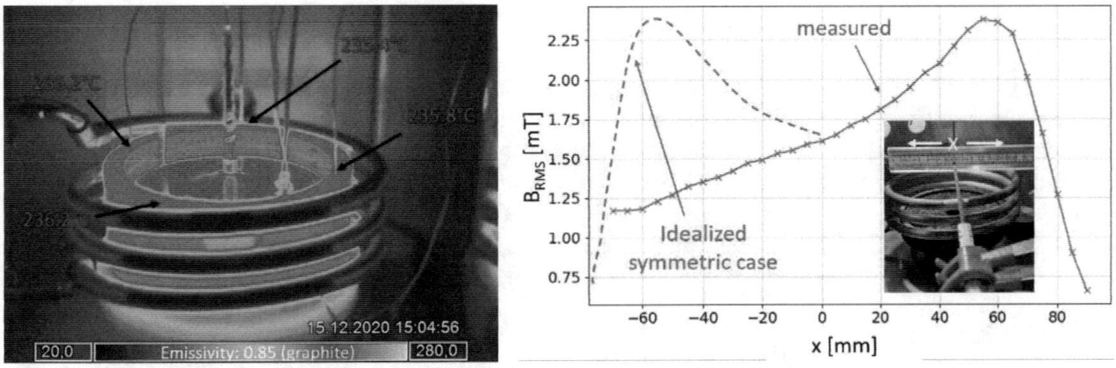

Figure 2. Experimental results for induction heating: thermal image during a growth process with TC measurements indicated by arrows (left); measurement of the vertical component of the magnetic field of the induction coil supplied with a current of 100 A (right).

An IR image from a growth experiment with the resistance heater is shown figure 3 (left, center). Only bottom insulation is used for better observability of the heater. Melt temperature is controlled to be 236°C. In contrast to the induction heating case, heat is produced in the heater volume from where it is transferred to the crucible by radiation. Therefore, the heater has an approximately 150°C higher temperature than the crucible.

Evaluation of heater temperature at different positions in figure 3 (left) indicates variations of 8 and 12°C between two adjacent curves of the meander and, respectively, two points in the center and at the edge of the same curve. Whether these variations are caused by local differences in heat generation, convective surface cooling or reflection artifacts caused by the steel furnace has to be investigated in future. At the inside of the heater, a temperature drop from 404 to 352°C is observed next to the power supply connection. In figure 3 (center) TC and Pt100 measurements of the crucible temperature are visualized, showing a circumferential asymmetry of 1.1°C. There is a tendency of the crucible to be colder towards the power supplies, however more accurate measurements are required for confirmation.

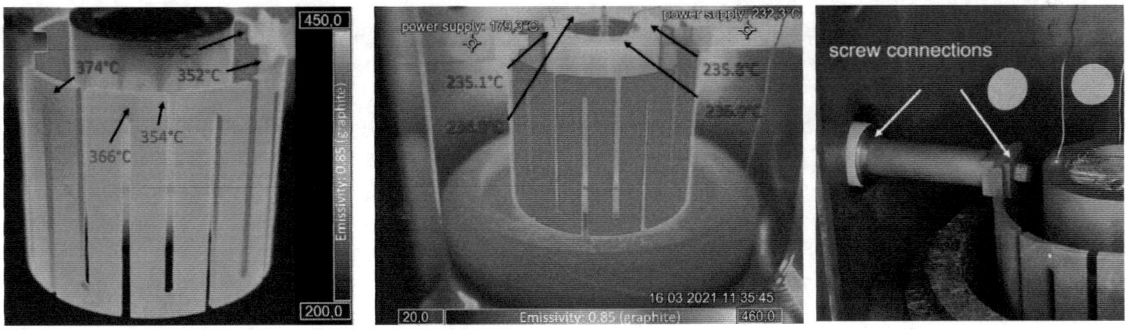

Figure 3. Thermal image of a growth experiment with resistance heater using bottom insulation (left, center) with IR camera (left) and TC measurements (center) indicated by arrows; optical image of the power supply (right).

The power supply temperature is evaluated in figure 3 (center) using the IR image in combination with fixed-emissivity stickers attached on the surrounding insulation. A difference of 53°C is observed, that in tendency fits to the crucible temperature asymmetry. This temperature difference is assumed to be caused by additional contact resistances at the screw connections shown in figure 3 (right). This is supported by a separate investigation in a table setup using a Sourcemeter device (Keithley, model 2450), where variation of the screw tightening at constant current of 1 A indicates large changes of the supply resistance in the range 28–163 mΩ. Additional voltage measurement directly at the graphite heater (excluding power supplies) and outside of the furnace (including power supplies) at different currents is summarized in table 1. The resulting power loss at the supplies is found to be decreasing with increasing current thus increasing temperature and is hence assumed to be caused by a temperature dependency of the contact resistances; additional investigation is required here. For the case with bottom insulation (95 A) shown in figure 3, 13% of the power is lost in the supplies.

Table 1. Measured electrical parameter (rms) with the graphite resistive heater.

Current [A]	U inside [V]	U outside [V]	P inside [W]	P outside [W]	Ratio [in/out]
19.9	2.2	2.7	43.8	53.7	0.81
40.5	4.0	5.0	162.0	202.5	0.80
60.5	5.7	6.8	344.9	411.4	0.84
80.1	7.1	8.3	568.7	664.8	0.86
100.2	8.5	9.8	851.7	982.0	0.87

The influence of the insulation level on the required heater power under air atmosphere is summarized in table 2; voltage measurements outside of the furnace are used for power determination. A power reduction of 64% is achieved comparing the experiments without and with full insulation; the cap has an influence of 19%. In a separate growth experiment without insulation under vacuum condition (see table 3) a 72% lower heater power is required, indicating a strong influence of convective cooling. Investigation of the influence of the insulation shape on the cooling airflow using a systematic comparison to the vacuum experiment could give further insights.

Table 2. Heater powers with different insulations (B=bottom, S=side, T=top) under air atmosphere.

without	B + ½ S	B + S	B + S + T
912.0 W	743.8 W	499.6 W	324.5 W

A qualitative comparison of heater power consumption in model experiments is summarized in table 3. The results show that, without insulation, the power consumption of the induction heating is lower than that of resistance heater by approx. 80% under vacuum and 90% under air. Comparing the fully insulated resistance heating setup from table 2 to the induction heating setup without insulation under air, there is still a difference of approx. 70%. However, it should be noted that these are the effective heater powers and not the actual powers consumed by the supply units. The power consumption of the induction generator could be significantly higher than the effective power of 45 / 105 W, while power losses in the supply of the resistance heater are expected to be much smaller [19]. These aspects will be investigated later.

Table 3. Comparison of heater powers from growth experiments with a control temperature of 236°C (without insulation). For the induction heater, the values are estimated by numerical simulation.

Heater	Vacuum	Air
Resistance	254.4 W	912.0 W
Induction	45.0 W	105.0 W

A general comparison between inductive and resistive heating is hardly possible: in crystal growth application there are various economic and process-specific requirements that are not considered in this study. For example, resistance heating generators work with DC or low frequency currents and are less complex and thus less expensive than induction heating generators at radio frequencies. However, inductively heated furnaces are better suited for growth processes with high temperatures (e.g. oxide crystals) or high purity requirements, because the crucible / melt are the hottest part in the system and no additional impurities are introduced at the heater surface.

4. Application to the crystal growth - experimental results

Growth experiments with both heater types are performed to analyze the achievable crystal diameter for various process parameters, cf. [12]. In the induction heating setup, the focus is set on the influence of pull velocity and melt control temperature under air atmosphere, vacuum and air cooling with an additional fan. The resulting diameters of 23 crystal grown under different conditions, evaluated at a crystal length of 100 mm, are shown in figure 4 (left); a selection of crystal photographs is presented in figure 4 (right). Generally, we observed decreasing crystal diameters with increasing melt temperature (crystals B, C, D, E) or pull velocity. This is the expected behavior from theoretical considerations. Active cooling of the crystal using the fan almost doubles the obtained crystal diameter (crystal A). In contrast, under vacuum conditions the resulting diameters are very small (crystal F). This indicates a strong influence of convective cooling in the present setup. On the surface of all crystals except F, a spiral structure is visible. The pitch of the spiral changes with pull velocity and crystal rotation (which is held constant here). The spiral structure is probably caused by an asymmetric temperature distribution. The spiral structure is most prominent on crystal A grown with asymmetric active cooling from the side.

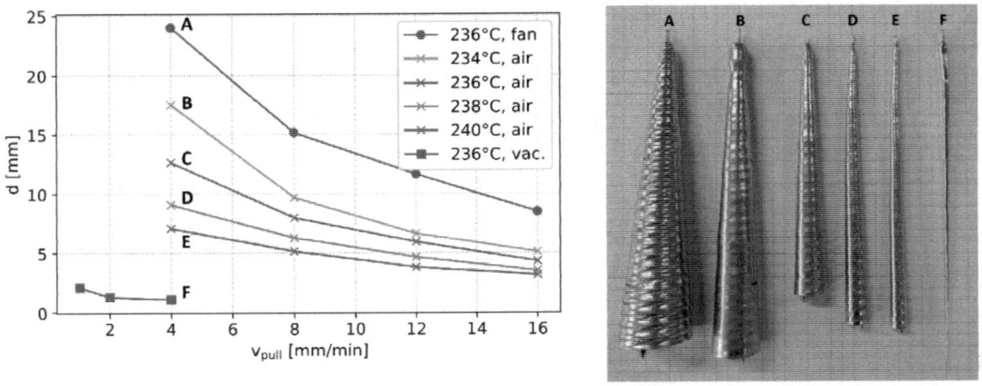

Figure 4. Experimental results from growth experiments with induction heating: resulting crystal diameters at length of 100 mm for different pull velocities, control temperatures and atmospheric conditions (left); crystals grown with a pull velocity of 4 mm/min (right).

In the resistance heating setup, the focus in the initial studies is set on the influence of the insulation. Air atmosphere and a melt control temperature of 236°C are used for investigation of the cases (a) without insulation, (b) bottom insulation, (c) half side insulation, and (d) full side insulation. In the cases (a) and (b) a growth velocity of 1 mm/min, in the cases (c) and (d) of 4 mm/min is used due to crystal size limitations. The resulting diameters and crystals are shown in figure 5. It is found that, in the present setup, increasing insulation increases the obtained crystal diameter. The assumed reason is that the heater temperature decreases with more insulation due to the lower power required to maintain the same melt temperature. In consequence, the radiative heating of the crystal is reduced. A spiral structure is visible on the surface of the crystals, similar to the inductively heated case.

In addition to the insights into the growth processes discussed above, the obtained results are highly valuable for validation of numerical simulations: broad parameter studies including comprehensive in-situ measurements are hardly possible outside the model experiment but are required for systematic examination of simplifications and assumptions made in modeling of crystal growth processes [7]. Phenomena observed in the present study, such as the convective cooling, asymmetries in temperature field, and asymmetric crystal shape are often neglected in literature but may have significant influence.

IOP Conf. Series: Materials Science and Engineering 1223 (2022) 012003 doi:10.1088/1757-899X/1223/1/012003

Further analysis with experiments and simulations optimized for scaling to real processes are planned for the future.

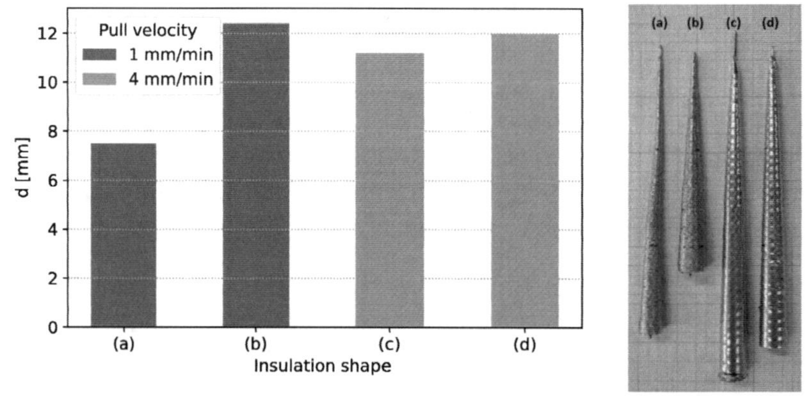

Figure 5. Experimental results from growth experiments with resistance heating: (a) without insulation, (b) bottom insulation, (c) half side insulation, and (d) full side insulation. Crystal diameter (left), obtained crystals (right).

5. Conclusions and outlook

A new model experiment setup for the CZ growth has been presented: extensive in-situ measurements are applied in a small-scale (melt diameter 50/100 mm) low-temperature process using the model material tin (melting point of 232°C) to get new insights into the characteristics of both induction and resistance heating during growth experiments.

For both induction and resistance heating, an asymmetric temperature distribution with deviations of approx. 1°C is observed in the crucible. In the induction heating setup, correlation with an asymmetric magnetic field of the inductor is assumed. In the resistance heating setup, heater temperature asymmetries are observed, particularly in the power supplies. At the power supplies of the resistance heater, temperature-dependent power losses of approx. 15% of the total heating power are measured, caused by contact resistances at screw connections. Despite nonexistent insulation, the induction heating setup outperforms the resistance heating in terms of effective heating power. However, the total power consumption of the supply units has not been investigated in detail so far. Further investigation is also planned regarding the asymmetric magnetic field of the inductor, the resistance heater temperature distribution, and the influence of the insulation shape on the convective cooling.

Growth experiments using the induction heating show an expected relation between melt temperature, pull velocity and crystal diameter; the obtained dataset gives valuable input for validation of numerical models [14,15]. In the resistance heating setup, the crystal diameter increases with increasing insulation level due to a reduction in heater temperature. Again, this behavior is as expected, but highlights the value of model experiments for investigation of design optimizations that is of much higher effort in real growth furnaces.

For future investigations, various new in-situ measurement techniques are currently under development, and a larger and more flexible experimental platform (MultiValidator) is under construction. Both will offer the basis for even more sophisticated experiments with in-situ measurements of the CZ process and other crystal growth techniques.

Acknowledgements
We thank the Technical Services team at IKZ for continuous support with experimental setups and lab infrastructure. The Test-CZ setup reused furnace parts from former projects managed by J. Wollweber at the IKZ. F.-M. Kießling (IKZ) is gratefully acknowledged for useful comments on the manuscript. This project has received funding from the European Research Council (ERC) under the European Union's Horizon 2020 research and innovation programme (grant agreement No 851768).

References
[1] Friedrich J, von Ammon W and Müller G 2015 Czochralski Growth of Silicon Crystals *Handbook of Crystal Growth* (Elsevier) pp 45–104
[2] Sarukura N, Nawata T, Ishibashi H, Ishii M and Fukuda T 2015 Czochralski Growth of Oxides and Fluorides *Handbook of Crystal Growth* (Elsevier) pp 131–168
[3] Ganschow S, Reiche P, Ziem M and Uecker R 2003 Controlling the axial temperature gradient in inductively heated Czochralski systems *Cryst. Res. Technol.* **38** 42–46
[4] Heidari H, Tavakoli M H, Shokri A, Moradi B M, Sharifi O M and Asaad M J M 2020 3D Simulation of the Coil Geometry Effect on the Induction Heating Process in Czochralski Crystal Growth System *Cryst. Res. Technol.* **55** 1900147
[5] Huang L Y, Lee P C, Hsieh C K, Hsu W C and Lan C W 2004 On the hot-zone design of Czochralski silicon growth for photovoltaic applications *J. Cryst. Growth* **261** 433–443
[6] Nam W, Hahn Y and Baik S 2017 Local optimization of graphite heater to save a power consumption of Czochralski Si ingot grower for PV application *Energy Procedia* **124** 767–776
[7] Dadzis K, Bönisch P, Sylla L and Richter T 2017 Validation, verification, and benchmarking of crystal growth simulations *J. Cryst. Growth* **474** 171–177
[8] Dadzis K, Pätzold O and Gerbeth G 2020 Model Experiments for Flow Phenomena in Crystal Growth *Cryst. Res. Technol.* **55** 1900096
[9] Nam P-O, Son S-S and Yi K-W 2010 The effect of polycrystalline rod insertion in a low Prandtl number melt for continuous Czochralski system *J. Cryst. Growth* **312** 1458–1462
[10] Krauze A, Jēkabsons N, Muižnieks A, Sabanskis A and Lācis U 2010 Applicability of LES turbulence modeling for CZ silicon crystal growth systems with traveling magnetic field *J. Cryst. Growth* **312** 3225–3234
[11] Bergfelds K, Pudzs M, Sabanskis A and Virbulis J 2019 Experimental and numerical investigation of laboratory crystal growth furnace for the development of model-based control of CZ process *J. Cryst. Growth* **522** 191–194
[12] Ekhult U and Carlberg T 1986 Czochralski growth of tin crystals under constant pull rate and IR diameter control *J. Cryst. Growth* **76** 317–322
[13] Schwabe D, Sumathi R R and Wilke H 2004 The interface inversion process during the Czochralski growth of high melting point oxides *J. Cryst. Growth* **265** 494–504
[14] Enders-Seidlitz A, Pal J and Dadzis K 2021 Experimental and numerical analysis of resistive and inductive heating in CZ growth, Proceedings of DKT 2021 Conference, 10.13140/RG.2.2.17384.19201
[15] Enders-Seidlitz A, Dadzis K and Pal J *J. Cryst. Growth* (to be submitted)
[16] Bernhard F 2014 *Handbuch der Technischen Temperaturmessung* (Berlin, Heidelberg: Springer Berlin Heidelberg)
[17] Vollmer M and Möllmann K-P 2018 *Infrared thermal imaging: fundamentals, research and applications* (Weinheim, Germany: Wiley-VCH Verlag GmbH & Co. KGaA)
[18] Available online: https://github.com/nemocrys/multilog
[19] Lupi S 2017 *Fundamentals of Electroheat* (Cham: Springer International Publishing)

IOP Conf. Series: Materials Science and Engineering 1223 (2022) 012004 doi:10.1088/1757-899X/1223/1/012004

Dynamic transitions of the magnetized spherical Couette flow between its base state and the return flow instability

J Ogbonna[1], F Garcia[2], T Gundrum[1], M Seilmayer[3] and F Stefani[1]

[1] Institute of Fluid Dynamics, Helmholtz-Zentrum Dresden-Rossendorf, Bautzner Landstr. 400, 01328 Dresden, Germany
[2] Universitat Politecnica de Catalunya, Escola d'Enginyeria de Barcelona Est, EEBE, Av. d'Eduard Maristany, 16, 08019 Barcelona, Spain
[3] SICK Engineering GmbH, Bergener Ring 27, 01458 Ottendorf-Okrilla, Germany

E-mail: j.ogbonna@hzdr.de

Abstract. The transition between the stable base state of the magnetized spherical Couette (MSC) flow and the return flow instability is experimentally investigated. The experiments are conducted using an MSC setup consisting of insulating spheres with the ratio of the inner to the outer radii $r_i/r_o = 0.5$, Reynolds number Re = 1000 and Hartmann number Ha $\in$ [25, 29]. The transition is characterized by changes in the power spectra of the azimuthal modes in the flow as Ha is dynamically changed. The transition occurs in the interval Ha $\in$ [26.5, 27.5]. The evolution of the power spectra of the azimuthal modes exhibits hysteretic effect depending on whether Ha is increased or decreased within the experimental interval. The power spectra in the azimuthal modes $m \in \{3, 4\}$ increases and remains dominant as Ha is increased, while the power spectra in $m \in \{2, 4\}$ are dominant while the flow is time dependent due to return flow instability as Ha is decreased.

1. Introduction

Spherical Couette flows are induced in fluids contained between two concentric spheres when the rotations of either one or both of the spheres cause the fluids to rotate differentially. If the fluid is electrically conducting and a magnetic field is applied, the resulting flow is referred to as a magnetized spherical Couette (MSC) flow. MSC flows are prevalent in astrophysical processes where ubiquitous amounts of plasmas and molten metals undergo differential rotation such as in the cores of the Sun, the planets and their moons.

MSC setups have been constructed as experimental devices to replicate various astrophysical processes in the laboratory to aid in their understanding. Experimental MSC setups are distinguished by the types of working fluids, the ratios of the radii of the inner to the outer spheres r_i/r_o, the electrical conductivities of the spheres and the configurations of the applied magnetic field. Liquid sodium with an electrically conducting inner sphere and an electrically insulating outer sphere has been used for MSC experiments with an axial magnetic field and $r_i/r_o = 0.33$ [1], and also with a dipolar magnetic field and $r_i/r_o = 0.35$ [2, 3, 4, 5]. The experiments in the present work are conducted with the HEDGEHOG (Hydromagnetic Experiment with Differentially Gyrating sphEres HOlding GaInSn) setup at Helmholtz-Zentrum Dresden-Rossendorf. HEDGEHOG uses GaInSn, a eutectic alloy of gallium (67% by mass), indium (20.5% by mass) and tin (12.5% by mass) as the working fluid. The ratio of the inner to the outer spheres $r_i/r_o = 0.5$, both spheres are electrically insulating, and an axial magnetic field is employed. Detailed descriptions of the experimental apparatus and the associated instrumentation are described in Section 3.

The simplest form of the MSC flow is where the inner sphere rotates while the outer sphere remains stationary. Such a flow is fully defined by four parameters, namely, the ratio of the radii of the inner to the outer spheres r_i/r_o, the Reynolds number Re = $\Omega r_i^2/\nu$ (where Ω and ν are the rotation speed of the inner sphere and the kinematic viscosity of the fluid, respectively), the magnetic Prandtl number Pm = ν/η (where η is the magnetic diffusivity of the fluid) and the Hartmann number Ha = $B_0 r_i/(\mu_0 \rho \nu \eta)^2$ = $B_0 r_i \sigma^{1/2}/(\rho\nu)^{1/2}$ (where B_0, μ_0, ρ and σ are the applied magnetic field, vacuum permeability, the density of the fluid and the electrical conductivity of the fluid, respectively). Four different states of the MSC flow are generally identified, one of which is steady and the other three are time dependent. The steady state of the MSC is its base state, which consists of a flow driven by the rotation of the inner sphere. This rotation flings the fluid outwards by inertia, thus creating a secondary meridional circulation consisting of a narrow jet along the equator towards the outer sphere, and a return flow from the rest of the outer sphere towards the origin of the jet [6].

The time-dependent states of the MSC flow are commonly referred to as the radial jet, the return flow and the shear layer instabilities. The stability curves demarcating the parameter space of Re and Ha into regions representing the base state and the primary time-dependent (periodic) states of the MSC flow have been numerically computed for r_i/r_o = 0.33 [6], 0.5 [7] and 0.6 [7]. Each curve represents the azimuthal mode of the flow as the curve is crossed from the base state to the periodic state. The use of bifurcation lines representing a particular property of the flow is also used to graphically represent the MSC flow states. Since the base state is purely axisymmetric, it is convenient to represent the kinetic energy of the non-axisymmetric azimuthal modes $m \neq 0$ with the bifurcation lines [8] to demarcate the base state from the periodic states. The radial jet instability is triggered by the rotation of the inner sphere, when the magnetic field is absent or low, as Re exceeds its critical value. As a result, the jet adopts a wavy structure, alternatively slightly above and below the equatorial plane [6]. The return flow instability arises in the presence of a moderate magnetic field, and the instability is characterized by the concentration of the kinetic energy in the return flow of the meridional circulation. A further increase in the magnetic field results in the shear layer instability, where the kinetic energy is concentrated in the shear layers formed along the surface axially projected by the inner sphere as a result of the increased magnetic field.

HEDGEHOG has been used to conduct experiments to investigate the transitions of the MSC flow between its base and time-dependent states [9], and more recently to investigate the return flow instability [10]. The former experiments were conducted at fixed values of Ha, while the latter experiments were conducted with dynamically changing values of Ha (but entirely within the confines of the parameter space for the return flow instability). The aim of the present work is to combine the essence of the previous experiments by investigating the transition between the base state and the return flow instability with dynamically changing values of Ha. In doing so, the range of Ha where its critical value for the transition occurs can be estimated. Furthermore, by increasing and decreasing the Ha within the transition range, the hysteresis effect of the multistability of the azimuthal modes may be observed just as during the previous investigation of the return flow instability [10]. Hence, we expect the present work will supplement the previous one [10] by investigating the onset of the return flow instability.

2. Equations

If the spheres in the MSC flow rotate about $\hat{\mathbf{e}}_z$, the unit vector representing the z-axis, the axial magnetic field applied externally to the flow is given by B_0

$$\mathbf{B}_0 = B_0\hat{\mathbf{e}}_z = B_0 \cos(\theta)\, \hat{\mathbf{e}}_r - B_0 \sin(\theta)\, \hat{\mathbf{e}}_\theta \tag{1}$$

where B_0, θ, $\hat{\mathbf{e}}_r$ and $\hat{\mathbf{e}}_\theta$ are the magnetic field, the colatitude and the unit vectors representing the radial distance and the colatitude, respectively. The fluid flow $\mathbf{v}$ is governed by the Navier-Stokes equation

$$\partial_t \mathbf{v} + \mathbf{v} \cdot \nabla \mathbf{v} = -\nabla p + \mathrm{Re}^{-1}\nabla^2 \mathbf{v} + \mathrm{Ha}^2\mathrm{Re}^{-1} (\nabla \times \mathbf{b}) \times \hat{\mathbf{e}}_z \tag{2}$$

with the no-slip condition imposed at the boundaries between the fluid and both spheres. **b** is the deviation of the magnetic field induced by the interaction of the fluid flow with $\mathbf{B}_0$. For Rm << 1, this induced magnetic field is negligible compared to $\mathbf{B}_0$. This so-called inductionless approximation applies here, since all experiments were conducted at Re = 1000 and the Pm of GaInSn is in the order of 10^{-6}, resulting in Rm in the order of 10^{-3}. The induction equation simplifies then to

$$0 = \nabla^2 \mathbf{b} + \nabla \times (\mathbf{v} \times \hat{\mathbf{e}}_z). \tag{3}$$

Insulating boundary conditions were considered in accordance with the experiments.

3. Experimental methods

3.1. Apparatus

The components of the HEDGEHOG setup are depicted in figure 1 (left). The acrylic glass inner and outer spheres have radii of 45 mm and 90 mm, respectively. The inner sphere is rotated by means of an electric motor shaft passing through the poles of both spheres. The diameter of the shaft inside the spheres is just 3 mm, so the interference which is very small relative to the interference on the overall flow of the fluid due to the rotation of the shaft is assumed to be insignificant. To act against the buoyancy force exerted by the relatively high density of GaInSn on the inner sphere, which is fully submerged in the fluid, a lead weight is contained within the inner sphere. The outer surface of the outer sphere contains provisions for mounting the ultrasonic transducers used for measuring the velocity of GaInSn. In addition, the surface contains 168 needle-like copper electrodes, providing the HEDGEHOG setup with the distinctive appearance that inspired its name. The copper electrodes enable the measurement of electric potential along the surface of the outer sphere if needed. The axial magnetic field for the flow is generated by a Helmholtz coil pair consisting of 80 wire windings in each coil. The distance between the coils and the radius of each coil are each approximately 300 mm.

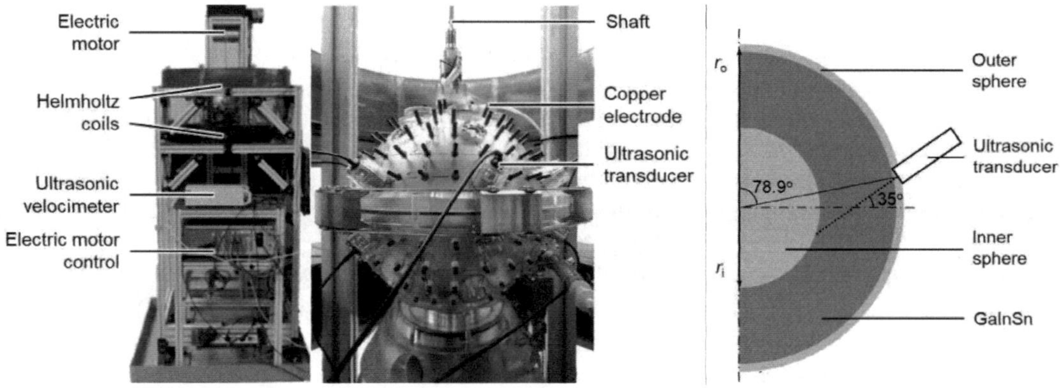

Figure 1. Left: photographs of the HEDGEHOG setup showing its full view and a close-up view on the outer sphere. Right: meridional section showing the geometric relations for one of the six azimuthally equidistant ultrasonic transducers installed on HEDGEHOG. The dotted line extending from the ultrasonic transducer represents its line of view along which the flow velocity components parallel to the line were measured.

3.2. Instrumentation

The ultrasonic Doppler velocimetry for measuring the velocity of GaInSn consists of piezoelectric ultrasonic transducers (Signal Processing SA, model TR0410LS) and an ultrasonic Doppler velocimeter (Signal Processing SA, model DOP3010). For the present experiments, six transducers were used at a colatitude of 78.9° with respect to the spheres. Each transducer was inclined southward so that the angle between the axis of the transducer and the equator of the spheres was 35°. Figure 1 (right) shows the geometric relationship between a transducer and the spheres. Each transducer periodically emits an ultrasonic pulse at a frequency of 4 MHz and received the echoes from the particles in the path of the ultrasonic beam. These particles are formed as a result of the oxidation of GaInSn due to its exposure to air [11]. From each transducer channel, the velocimeter sequentially acquires the echo signals reflected by the particles to determine the velocities of the flow parallel to each transducer axis.

4. Results

Since ultrasonic Doppler velocimetry relies on the reflection of the ultrasonic pulses by the naturally-occurring oxides in GaInSn, the oxides were distributed by circulating the GaInSn from the north pole to the south pole via a peristaltic pump connected to the poles externally. The values of Re and Ha for the experiments were set using the properties of GaInSn listed in table 1. All experiments were conducted at Re = 1000. The magnetic field B_0 for the required Ha was set by the current I through the Helmholtz coil according to the relation

$$B_0 = (4/5)^{3/2} \mu_0 nI/R \tag{4}$$

where n is the number of wire windings in each Helmholtz coil and R is the radius of each coil and the distance between the coils.

Table 1. Physical properties of GaInSn from [12].

ρ (kg m^{-3})	ν (m^2 s^{-1})	$\eta = 1/\mu_0\sigma$ (m^2 s^{-1})
6.4×10^3	3.4×10^{-7}	0.2428

We first established the range of Ha within which the experiments were conducted. From numerical computations, for $r_i/r_o = 0.5$ and Re = 1000, the critical value of Ha at which the MSC flow transitions between the base state and the return flow instability is 25.8 [7]. Below this critical value, the MSC flow is in its base state but above the critical value, the flow becomes time dependent due to return flow instability. Previous experiments with HEDGEHOG showed that at Re = 1000 and Ha = 25, the flow was in its base state. Hence, the critical value of Ha for the flow in HEDGEHOG is at least 25, and this value was selected as the lower Ha bound for the experiments. Results from [10] showed that at Re = 1000, the flow was time dependent due to return flow instability at Ha = 27.5. This indicates that the critical value is below Ha = 27.5. However, in order to minimize the effect of uncertainties in the values Ha, the experimental range was widened by selecting Ha = 29 as the upper bound for the experiments. Therefore, the experiments were conducted at Ha $\in$ [25, 29] and the values of Ha were increased and decreased (linearly with time) between the limits. Each experiment was conducted for a period of 6 hours. In the subsequent sections, we first discuss the velocity profile of the flow during the experiments, and afterwards, the evolution of the azimuthal modes in the flow.

4.1. Velocities

Figure 2 and figure 3 show the plots of the velocity components of the flow along the axis of one of the six ultrasonic transducers as Ha was respectively increased and decreased between 25 and 29 (see top horizontal axes). The depths indicate the distance away from the transducer along its axis. The sign convention of the velocity is such that flows away from the transducers have positive velocities, while flows towards the transducers have negative velocities. Looking at figure 2, we see that the closer to the transducer, the flow has velocities that are nearly zero in value, while the flow velocities are negative further away from the transducer. The near-zero velocities close to the transducer are those of the return flow of the meridional circulation of the fluid. Since this flow is along the surface of the outer sphere, it cuts across the transducer axes nearly perpendicularly, thus registering velocities that are nearly zero. The negative velocities further away from the transducer are those of the radial jet of the meridional circulation. Since the radial jet originates from the inner sphere and flows towards the outer sphere, and hence the transducers, it registers as having negative velocities. As Ha increases from 25 to 29, the influence of the return flow in the meridional circulation increases while diminishing the influence of the radial jet.

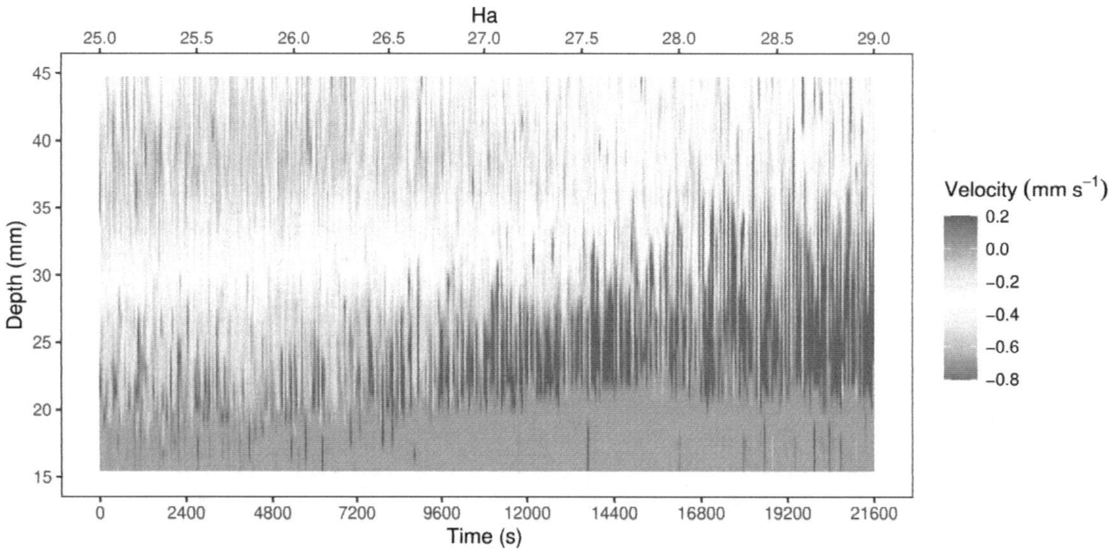

Figure 2. Plots of the experimentally measured velocity components parallel to the ultrasonic transducer axis as Ha was increased from 25 to 29. The depth indicates the axial distance from the ultrasonic transducer. Re = 1000.

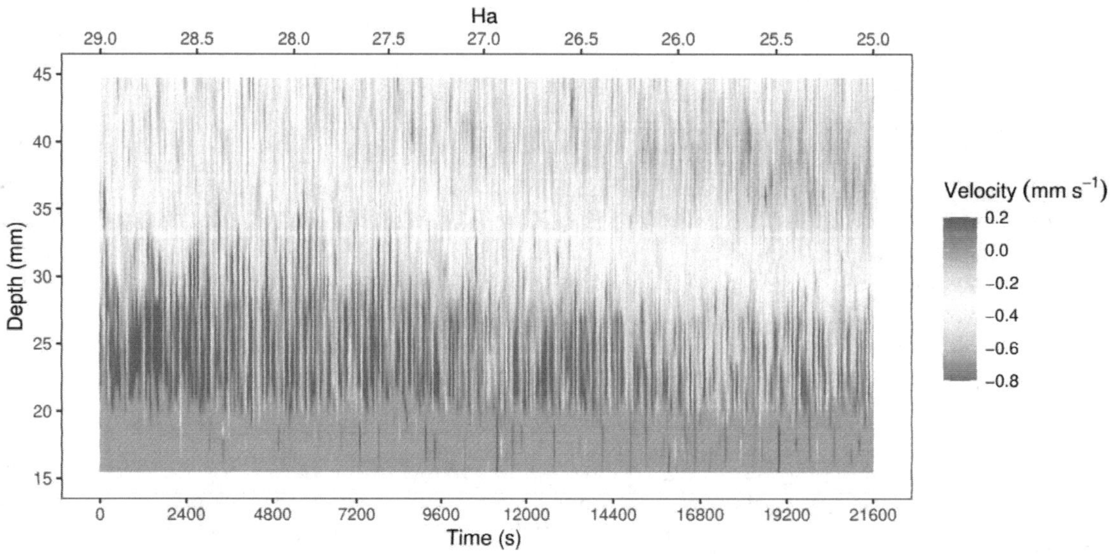

Figure 3. Plots of the experimentally measured velocity components parallel to the ultrasonic transducer axis as Ha was decreased from 25 to 29. The depth indicates the axial distance from the ultrasonic transducer. Re = 1000.

Since figure 2 and figure 3 were conducted within the same limits albeit from opposite ends, it is not surprising that the two plots resemble mirror images of each other. Consequently, earlier analyses of figure 2 applies to figure 3.

4.2. Azimuthal modes

It was mentioned in Section 1 that the base and time dependent states of the MSC flow differ by the absence of kinetic energy in the non-axisymmetric azimuthal modes in the former. Since the kinetic energy correlates with the power spectra of the azimuthal modes obtained from the Direct Fourier Transform (DFT) of the velocity-time data of the flow, the critical value of Ha for the flow transition in the experiments were estimated by the change in the power spectra of the azimuthal modes. The velocity-time data were divided into rectangular windows of 1080 seconds or Ha intervals of 0.2. Within each window, a two-dimensional DFT of the data in time and space (given by the positions of the six ultrasonic transducers) was conducted at a distance of 30 mm from the transducers, corresponding to coordinates of $(r, \theta) = (64 \text{ mm}, 90°)$. The resulting amplitudes from the DFT were first squared, and then divided by twice the variance of the velocities in each window to obtain the normalized power spectra. The points in figure 4 and figure 5 indicate the root sum squares of the normalized power spectra in each azimuthal mode as Ha was respectively increased and decreased between 25 and 29. The curves indicate the locally estimated scatterplot smoothing (LOESS) of the points with a polynomial degree of 2 and a span of 0.5.

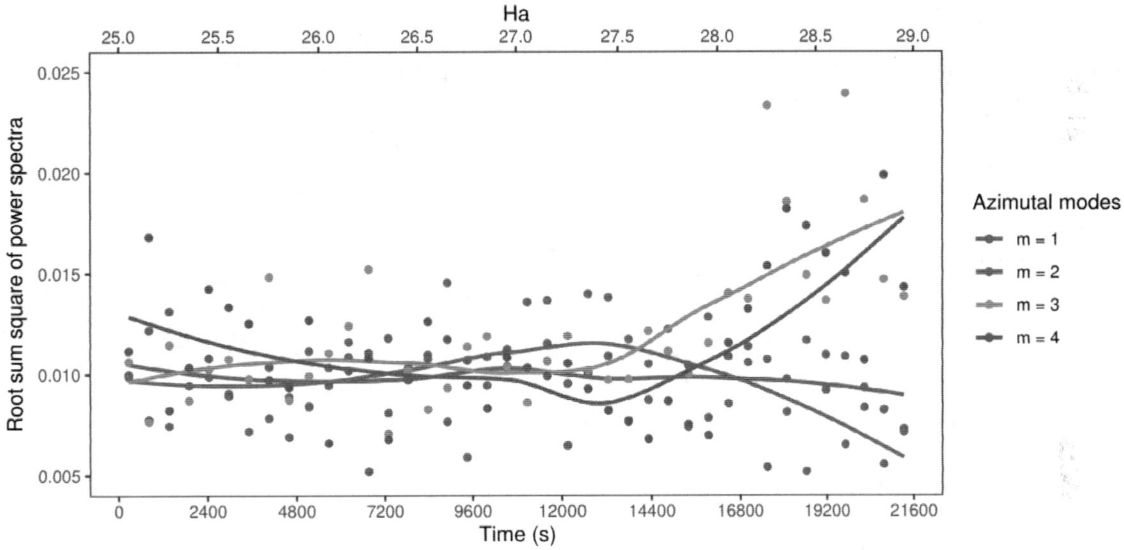

Figure 4. Root sum squares of the normalized power spectra in each azimuthal mode obtained from the DFT of the velocity-time data as Ha was increased from 25 to 29. Re = 1000.

Figure 4 shows that the power spectra of the azimuthal modes $m \in \{3, 4\}$ begin to increase within Ha $\in$ [27.0, 27.5]. Meanwhile, figure 5 shows that the power spectra in $m \in \{2, 4\}$ initially increase, as the flow increasingly becomes saturated, and then begins to decrease from Ha =27.5 until they fall to values similar to the rest of the azimuthal modes at around Ha = 27.5. Hence, the critical point of transition between the base state and the return flow instability for the experiments is in the interval Ha $\in$ [26.5, 27.5]. Given that the numerically computed critical value is Ha = 25.8, the discrepancy from the experimental result is approximately 3% to 7%.

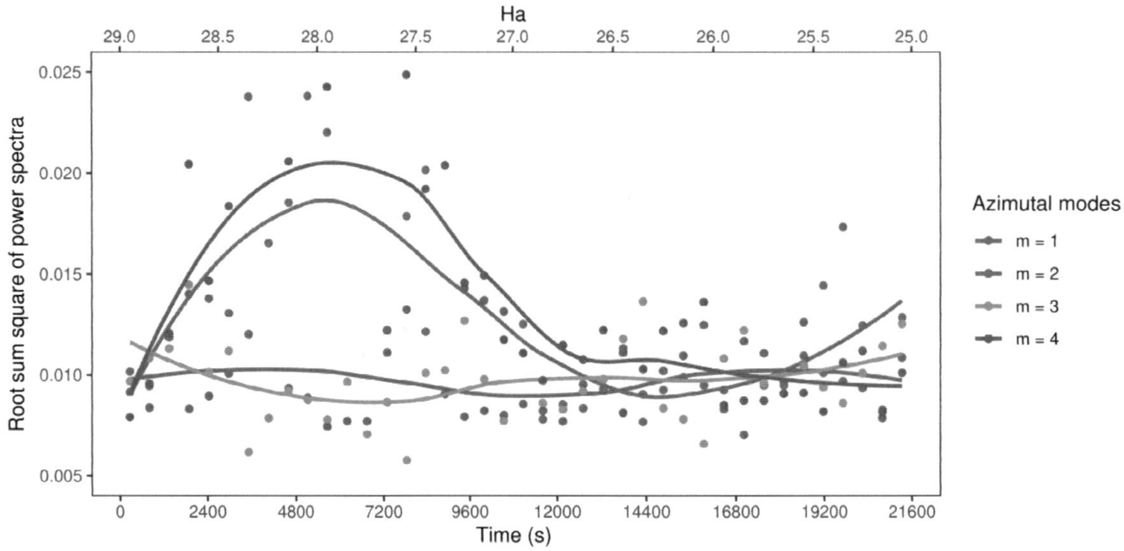

Figure 5. Root sum squares of the normalized power spectra in each azimuthal mode obtained from the DFT of the velocity-time data as Ha was decreased from 25 to 29. Re = 1000.

5. Conclusion

Experiments were conducted as the values of Ha were dynamically increased and decreased within the interval [25, 29] at Re = 1000, spanning the base state and the time dependent state due to return flow instability of the MSC flow. This was done in order to estimate the critical value of Ha for the transition and to observe the behavior of the system. The transition was characterized by the change in the power spectra (and hence, the kinetic energy) of the azimuthal modes in the MSC flow and was observed to occur within Ha ∈ [26.5, 27.5]. The evolution of the power spectra of the azimuthal modes exhibited hysteretic effects, as the power spectra in the azimuthal modes $m \in \{3, 4\}$ increased and remained dominant as Ha increased within the experimental interval, while power spectra in $m \in \{2, 4\}$ were dominant while the flow was time dependent due to return flow instability.

The azimuthal modes of the MSC flow correspond to the azimuthal symmetry of the waves in the flow. The simplest form of these waves are called rotating waves, which have been investigated for the return flow instability in [10]. A wave with one frequency – that of modulation – in addition to its frequency of rotation is referred to as a modulated rotating wave. Waves with two and even three additional frequencies have also been numerically computed [13]. Solutions to modulated rotating waves have been found through numerical simulations in the return flow instability, but since the range of Ha in which they the return flow instability is extremely narrow [14], detecting them experimentally is a challenging exercise. Fortunately, the range of Ha for the occurrence of modulated rotating waves in the radial jet instability is much wider [15] and it is very likely that they can be detected experimentally. Consequently, it is worthwhile to experimentally investigate the radial jet instability in detail.

6. References

[1] Sisan D R, Mujica N, Tillotson W A, Huang Y M, Dorland W, Hassam A B, Antonsen T M and Lathrop D P 2004 *Phys. Rev. Lett.* **93** 114502

[2] Schmitt D, Alboussière T, Brito D, Cardin P, Gagnière N, Jault D and Nataf H C 2008 *J. Fluid Mech.* **604** 175–197

[3] Nataf H C, Alboussiere T, Brito D, Cardin P, Gagnière N, Jault D and Schmitt D 2008 *Phys. Earth Planet. Inter.* **170** 60–72

[4] Brito D, Alboussière T, Cardin P, Gagnière N, Jault D, La Rizza P, Masson J P, Nataf H C and Schmitt D 2011 *Phys. Rev. E* **83** 066310

[5] Figueroa A, Schaeffer N, Nataf H C and Schmitt D 2013 *J. Fluid Mech.* **716** 445–469

[6] Hollerbach R 2009 *Proc. R. Soc. A* **465** 2003–2013
[7] Travnikov V, Eckert K and Odenbach S 2011 *Acta Mech.* **219** 255–268
[8] Garcia F and Stefani F 2018 *Proc. R. Soc. A* **474** 20180281
[9] Kasprzyk C, Kaplan E, Seilmayer M and Stefani F 2017 *Magnetohydrodynamics* **53** 393–401
[10] Ogbonna J, Garcia F, Gundrum T, Seilmayer M and Stefani F 2020 *Phys. Fluids* **32** 124119
[11] Borin D, Nikrityuk P and Odenbach S 2009 *Appl. Rheol.* **19** 61995
[12] Rüdiger G, Kitchatinov L L and Hollerbach R 2013 *Magnetic processes in astrophysics: theory, simulations, experiments* ISBN 978-3-527-41034-7
[13] Garcia F, Seilmayer M, Giesecke A and Stefani F 2020 *Chaos* **30** 043116
[14] Garcia F, Seilmayer M, Giesecke A and Stefani F 2019 *J. Nonlinear Sci.* **29** 2735–2759
[15] Garcia F, Seilmayer M, Giesecke A and Stefani F 2020 *Phys. Rev. Lett.* **125** 264501

Acknowledgments
This project has received funding from the European Research Council (ERC) under the European Union's Horizon 2020 research and innovation programme (Grant Agreement No. 787544).

IOP Conf. Series: Materials Science and Engineering 1223 (2022) 012005 doi:10.1088/1757-899X/1223/1/012005

Design of induction tempering of surface hardened components

Marco Baldan, Bernard Nacke

Institute of Electrotechnology, Leibniz University Hannover, Wilhelm-Busch-Strasse 4, 30167, Hannover, Germany

E-mail: baldan@etp.uni-hannover.de

Abstract. Induction hardening is a well-known heat treatment in the automotive and aerospace industries. Very often it is accompanied by tempering that is conventionally performed in ovens. However, induction tempering is getting more popular and this introduces the chance of using numerical simulations and optimization in the design stage. In this paper, based on a real case study, we will describe how to model and choose the regime parameters in both induction hardening and tempering.

1. Introduction

Mechanical parts, such as shafts, pinions, gears, and hubs, are ever-present components in automotive and aerospace. They are subjected to surface treatments in order to improve their ware behaviour and vibration resistance. Induction hardening determines a hard surface layer with compressive stresses, which has proved to be effective in augmenting the piece's fatigue life and wear resistance [1]. After hardening, the components are often tempered to diminish the chance of crack formation and to enhance toughness and ductility. The interest in induction tempering has grown in the last years, since in comparison to the conventional treatment (i.e. in oven), the process time is significantly reduced (seconds instead of hours) [2-6]. Moreover, induction tempering is easily embeddable in the production line.

In 2019, almost 92M (i.e. 92 millions) vehicles were produced worldwide and this trend is growing. The world of automotive is experiencing a significant change that will strongly affect the entire sector of induction heating (machine producers, hardening shops...) too. It is mostly probable that the amount of inductively treated pieces will be heading for a contraction in the nearest future. But nonetheless, the induction technology will be pushed to achieve better standards and quality. Using numerical modelling can definitely help in this sense. Multi-physical numerical modelling of induction hardening has a long story that started in 90s [7-8]. The interest in simulating tempering is much more recent. Several works regarding multi-physical modelling of the conventional treatment do exist [9-12]. When one refers instead to induction tempering, models are typically limited [4], [6].

In this paper, a multi-physical numerical model of both hardening and tempering was adopted. The model includes the electromagnetic, thermal, and metallurgical analyses. This is already a novelty. Moreover, our work focuses on the design of the processes too. So far, the use of multi-physical models in induction tempering has not been extensively investigated. The paper is developed as follows. Firstly, a quick description of the numerical model is given. In a separate work we will describe the model more in detail. Then, different strategies regarding how to design the regimes in the induction hardening-tempering process are investigated.

2. Multi-physical numerical model

The multi-physical numerical model of induction hardening-tempering includes three different physical analyses. The electromagnetic analysis is described by the Maxwell equations evaluated in frequency domain [14]:

$$\nabla \times (\mu_r \mu_0)^{-1} \nabla \times \boldsymbol{A} + j\omega\, (\mu_r \mu_0)\, \rho^{-1} \boldsymbol{A} = -\rho^{-1}\, \nabla V\,, \tag{1}$$

$$\nabla \cdot \rho^{-1} (j\omega \mu_r \mu_0\, \boldsymbol{A} + \nabla V) = 0, \tag{2}$$

where A stands for the magnetic vector potential, V represents the electric scalar potential, $j = (-1)^{0,5}$, and ω is the angular frequency. The magnetic permeability (μ) results to be not only temperature and field strength, but also microstructure dependent [13]. The electrical resistivity (ρ) is temperature and microstructure dependent as well. The thermal analysis relies on the heat conduction equation:

$$\nabla \cdot (\lambda\, \nabla T) + \rho_t c_p \frac{\partial T}{\partial t} = \dot{q} \tag{3}$$

In which T denotes the temperature, λ indicates the thermal conductivity, ρ_t and c_p the density and the specific heat, respectively. $\dot{q}$ describes the heat source density.
The microstructural analysis involves both heating and quenching stages in case of hardening. Therefore it is able to predict the distribution of austenite and martensite. In case of tempering, the model evaluates the map of carbides and the final hardness (H). This is obtained as a rule of mixture from the different microstructures. In particular, the hardness of tempered martensite is calculated as a linear dependency from the Hollomon-Jaffe tempering parameter [15]:

$$TP = \frac{T+273.15}{1000}\left(C + \log\frac{t}{3600}\right) \tag{4}$$

It is straightforward to recognize that, compared to the conventional treatment, induction tempering, due to the short heating time, requires higher temperatures to get an equivalent hardness reduction. However, the temperature increase is much more limited in comparison to the time diminution.

3. Optimal design of the hardening process

Designing a hardening process involves mainly two aspects. On the one hand, the definition of the optimal regime. This includes the heating time, the frequency and the power (or inductor current, or voltage). On the other hand, there is the inductor geometry. One possible way to find the optimal parameters (i.e. current, frequency, time, shape and dimensions of the inductor) is to build a numerical model, formulate an objective function, which is dependent on all these parameters, and couple the model with some optimization algorithm. However, a smarter approach does exist [16]. Indeed, the regime parameters can mostly influence the hardness (i.e. martensite) distribution in the depth. On the contrary, the inductor geometry influences the contour of the hardness profile. To a first approximation, since the desired hardening depth (hd) is known when designing a hardening process, the regime parameters can be adjusted in such a way that the wished hardening depth is achieved. In order to do that, a 1D model could be employed to define the optimal time and frequency, without resorting to using the numerical model with the real geometry. From this analysis, an estimation of the needed density power is also obtainable. Once time and frequency have been set, the process design can be focused only on the inductor geometry. The power (or inductor current or voltage) needs only to be adjusted. Therefore, one can think about optimizing the inductor through an optimization problem in which only geometrical variables appear (in addition to the adjustment of power or current).

IOP Conf. Series: Materials Science and Engineering 1223 (2022) 012005 doi:10.1088/1757-899X/1223/1/012005

In the 1D model one can recognize three parts: the inductor (a massive piece of copper), the air gap, and the piece (made of steel). The task is to find the optimal regime, given the desired hardness depth. For simplicity, it is assumed that the austenite and martensite distributions coincide. This is reasonable for thin (few mm) hardness depths and relevant quenching intensities. Moreover, the full austenitization model will be not used. The aim is that the whole hardness depth overcomes the A_{C3} temperature. It is interesting to study the trade-off between the surface temperature and energy consumption. This is formulated as:

$$\min_{I,f,t_h} \boldsymbol{F}(I,f,t_h) \ with\ I = \arg\min_{i} g(i,t,t_h) \tag{5}$$

in which I is the current (assumed constant during the heating), f is frequency, and t_h is the heating time. The objective functions are $f_1 = T(t_h, 0)$ (the surface temperature at the end of heating), $f_2 = \int P\,dt$ (the energy consumption), and $g = |T(t_h, hd) - A_{C3}|$, which represents the deviation from A_{C3} of the final temperature at the border of the hardening depth. In this case, the frequency range is [5-30] kHz, while the heating time changes between 1 and 5 s. Once (5) has been solved, it turns out that the "image" of the non-dominated solutions shows that, surprisingly, the front can be divided into two parts. In the "left" part, all the points assume the minimum range frequency, whereas in the "right" one, the heating time is always at minimum. From these results one can conclude that the frequency plays a minor role compared to the heating time in terms of energy consumption. In other words, in this approach, the choice of the variable ranges directly affects the obtained Pareto front. Therefore, a new strategy was adopted. In the Pareto front, the point with the lowest energy consumption is the point with the optimal frequency and minimum time. This means that a multi-objective optimization approach is here overabundant. Considering a typical heating time, the optimal frequency could be evaluated considering the energy consumption (f_2) as a unique objective function. With a heating time of 1 s, the minimum is located between 19 and 20 kHz (Fig. 1). The curve is a bit noisy because the value of g is not always zero and it differs for each individual.

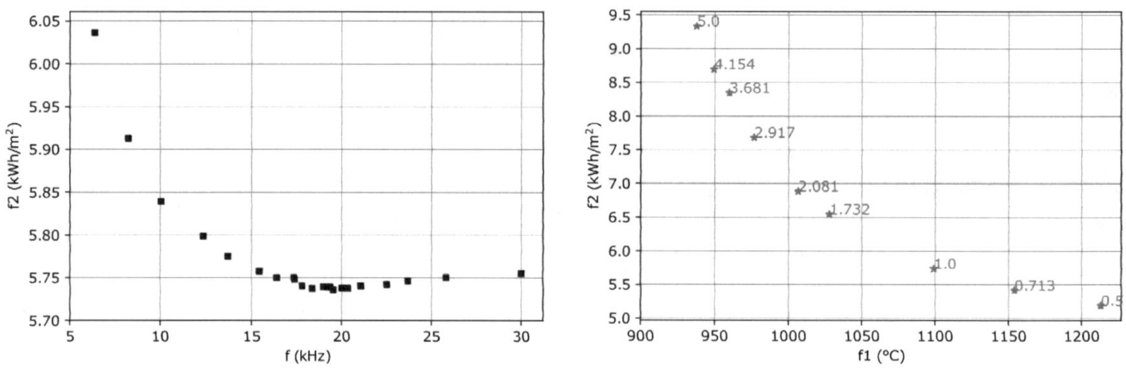

Figure 1: Energy consumption as a function of frequency (t_h=1 s, hd=2 mm) (a); trade-off between surface temperature and energy consumption with f = 19.5 kHz. Labels indicate the heating time (b).

In case of a 2.2 mm hardening depth, with a heating time of 0.5 s, the optimal frequency is 15.2 kHz (theoretical value is 12.4 kHz [17]). This motivates the choice of working with a frequency in between in the real application (Fig. 2). The geometry of the one-turn inductor is assumed given, the component is a L-profile made of AISI 4140. The hardness level after hardening is approximately 62 HRC.

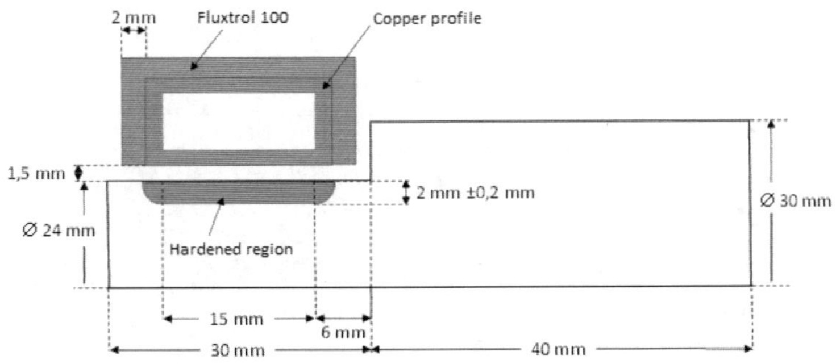

Figure 2: Sketch of the real application. A one-turn inductor with magnetic flux concentrator is used.

4. Optimal design of the tempering process

In this section the tempering regime is optimized, in order to smooth the hardness distribution around a value within the hardness depth. The inductor geometry will not change (Fig. 2). It means that the hardness in the axial direction is only slightly affected. The task is an optimal control problem. Inspired by the "alternance method" approach [18], the study starts with only one pulse followed by the soaking time (2 variables) and up to three pulses will be considered. t_a denotes the time in which the power supply is on, t_s when it is off. After the last soaking stage, the piece is quenched down. The inverse problem is defined as the maximum deviation from 58 HRC within the hardening depth:

$$\arg \min_{t_{a1},\dots,t_{sN}} \max |H - 58| \tag{6}$$

Therefore, we would like to reduce of 4 HRC the hardness level, which, in case of induction tempering, represents a challenging task. The power is the minimum achievable in the real application and it is equal to 8 kW. No constraints relative to the tempering stage or to the values of TP are implemented. Tab. 1 indicates the obtained values of the optimal variables. As expectable, increasing the number of pulses allows to flatten more and more the hardness distribution around 58 HRC. Tab. 1 also recalls the hardness difference within the hardness depth. Notice that the maximum deviation from 58 HRC is approximately the half of the hardness difference ΔH. This is a good sign. Alternance method makes use of this property to define in advance how the optimal solution is supposed to look like. Fig. 3 shows the three optimal hardness distributions. Only in case of three pulses a successful final hardness profile is obtained.

Table 1. Results of the three inverse problems relative to the heating regime and hardness distribution. The unit of time is second, the one of hardness HRC.

Inverse Problem IP	Pulses	t_{a1}	t_{s1}	t_{a2}	t_{s2}	t_{a3}	t_{s3}	ΔH	$\max\|H - 58\|$
1	1	1.69	4.64	-	-	-	-	6.40	3.21
2	2	1.41	1.24	0.60	1.91	-	-	4.94	2.47
3	3	1.27	3.65	0.68	1.67	0.47	2.40	4.13	2.09

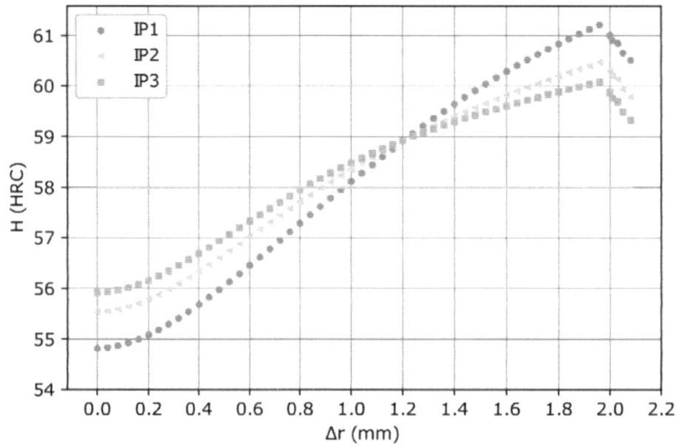

Figure 3. Optimized hardness distribution in the three inverse problems.

From the previous analysis one can conclude that induction tempering is particularly suitable for small hardness reductions. In case the desired hardness is 58 HRC, the reached maximum deviation was never below 2.09 HRC. In general, in industrial applications, a tolerance of ±2 HRC is the standard. Now, still referring to the same setup (piece, inductor, frequency), the aim is trying to further smooth the hardness distribution after tempering around 58 HRC. So far, the quenching process after hardening was 15 s long and it allowed to cool the piece up to room temperature. However, the formation of martensite is almost completed around 200 °C (at least in case of AISI 4140). By shortening the quenching time (t_Q), it is possible to use the heat still present in the piece (especially close to the core) to get a tempering effect on the martensitic surface (self-tempering).

It is now investigated whether combining self- with induction tempering could be beneficial in order to get a final homogeneous hardness of 58 HRC. It is still assumed that tempering is performed right after hardening with same power supply, and therefore the minimum power available is 8 kW. Two cases are considered: in the first one, the quenching time is 3 s, in the second is 2.5 s. Afterwards, two inverse problems are run (IP4, IP5), aiming at smoothing the hardness curve around 58 HRC. Heating regimes are defined by two pulses. Hardness results are depicted in Tab. 2. Having a quenching time of 3 s does not seem to be so helpful. With a successive two pulses tempering, the maximum deviation from 58 HRC is 1.91 HRC (it was 2.47 HRC in the case without self-tempering). On the contrary, when the quenching time is 2.5 s, again with a subsequent two steps tempering, the max-norm falls to 1.22 HRC. This leads to a significant hardness flattening.

Table 2. Hardness distribution after self-tempering and optimized induction tempering (8 kW).

Inverse Problem IP	Pulses	t_Q	t_{a1}	t_{s1}	t_{a2}	t_{s2}	ΔH	$\max\lvert H - 58\rvert$
4	2	2.5	1.42	3.10	0.61	2.71	3.77	1.91
5	2	3	1.32	2.31	0.51	3.71	2.40	1.22

5. Conclusions

In this paper it has been introduced a multi-physical model of both induction hardening and tempering. However, the main focus has been on the design of the processes. Firstly, it has been described how to choose the optimal heating time and frequency in hardening. Then, it has been investigated how to flatten the hardness profile around 58 HRC using tempering. Combining self- and induction tempering resulted to be a winning strategy, since the obtained maximum deviation was only 1.2 HRC.

6. References

[1] Rudnev V, Loveless D, Cook D and Black M 2003 *Handbook of induction heating* Marcel Dekker Inc.

[2] Euser V, Williamson D, Clarke K D, Findley K O, Speer J G and Clarke A J 2019 Effects of Short-Time Tempering on Impact Toughness, Strength, and Phase Evolution of 4340 Steel Within the Tempered Martensite Embrittlement Regime *Metallurgical and materials Transactions A* **50**(A) pp. 3655-3662.

[3] Kaiser D, De Graaff B, Dietrich S and Schulze V 2017 Investigation of the tempering process of martensitic AISI 4140 steel at high heating rates *Proc. of IFHTSE Congress 2017*.

[4] Zabett A and Azghandi S 2012 Simulation of induction tempering process of carbon steel using finite element method *Materials and Design* **36** pp. 415-420.

[5] Sackl S, Zuber M, Clemens H and Primig S 2016 Induction Tempering vs Conventional Tempering of a Heat-Treatable Steel *Metallurgical and Mat. Tra. A* **47**A pp. 3694-3702.

[6] Vieweg A, Raninger P, Prevedel P, Ressel G, Ecker W, Marsoner S and Ebner R 2017 Experimentelle und numerische Untersuchung des induktiven Anlassens eines Vergütungsstahles *HTM J. Heat Treatm. Mat* **4** pp. 199-204.

[7] Bokota A and Iskierka S 1998 Numerical analysis of transformations and residual stresses in steel cone-shaped elements hardened by induction and flame methods *Int. J. Mech. Sci* **40**(6) pp. 617-629.

[8] Simsir C 2014 Modeling and Simulation of Steel Heat Treatment-Prediction of Microstructure, Distortion, Residual Stresses, and Cracking *ASM Handbook* **4**B pp. 409-466.

[9] Deng X and Ju D 2013 Modeling and Simulation of Quenching and Tempering Process in Steels *Physics Procedia* **50** pp. 368-374.

[10] Eser A, Broeckmann C and Simsir C 2016 Multiscale modeling of tempering of AISI H13 hot-work tool steel – Part 2: Coupling predicted mechanical properties with FEM simulations *Computational Materials Science* **113** pp. 292-300.

[11] Tong D, Gu J and Yang F 2018 Numerical simulation on induction heat treatment process of a shaft part: Involving induction hardening and tempering *J. of Materials Proces. Tech.* **262**.

[12] Baldan M, Nikanorov A and Nacke B 2020 A parallel multi-fidelity optimization approach in induction hardening *COMPEL - The international journal for computation and mathematics in electrical and electronic engineering* **39**(1) pp. 133-143.

[13] Baldan M, Stolte M H, Nacke B and Nuernberger F 2020 Improving the Accuracy of FE Simulations of Induction Tempering Toward a Microstructure-Dependent Electromagnetic Model *IEEE Transactions on Magnetics* **56**(10).

[14] Schlesselmann D 2016 *Methoden zur numerischen Berechnung, Analyse und Auslegung induktiver Randschichthärteprozesse unter Berücksichtigung von magnetischen Sättigungseffekten* PZH Verlag.

[15] Hollomon J and Jaffe L 1945 Time-temperature relations in tempering steel *Trans. AIME* 162:223.

[16] Baldan M, Nikanorov A and Nacke B 2018 Hierarchical optimization approaches in designing surface hardening induction systems *IOP Conf. Ser.: Mater. Sci. Eng.* 424 012067.

[17] Sluhotsky A and Ryskin C 1974 *Inductors for the induction heating* Energia.

[18] Rapoport E and Pleshivtseva Y 2006 *Optimal control of induction heating processes* T&F.

Acknowledgments
Authors would like to acknowledge the program for the promotion of cooperative industrial research (IGF No. 20008 N) via the German Federation of Industrial Research Associations (AiF) e. V. on the basis of a resolution of the German Bundestag.

IOP Conf. Series: Materials Science and Engineering 1223 (2022) 012006 doi:10.1088/1757-899X/1223/1/012006

Twenty Years' Research on Microwave Application to Metal Production and Recycling

Noboru Yoshikawa

6-6-02, aza-Aoba, Aramaki, Aoba-ku, Sendai, Japan980-8579
Department of Materials Science and Engineering, Graduate School of Engineering, Tohoku University

E-mail: yoshin@material.tohoku.ac.jp

Abstract. Ever since discovery of microwave heating in 1946, its application has been developed in various fields. Our research group has performed intensive studies on microwave heating application to metal production and recycling for these twenty years. In this article, it is intended to introduce our attempts having made for different projects. They are 1. Microwave processing of Ti bearing blast furnace slag, microstructural alteration and comminution. 2. Microwave carbo-thermic reduction of valuable metals from various oxides. 3. Microwave drying and dehydration of wet wastes. 4. Vapor de-phosphorization by microwave rapid heating. It is to be emphasized that we utilized a single mode microwave applicator for the purpose of clarification on the heating mechanisms of various metallurgical matters and compounds, and it was intended to discuss the reaction kinetics excited by microwave heating.

1. Introduction

Microwave heating has been applied to many fields since the first patent of microwave heating is issued in 1945 [1]. The major application might be the domestic ovens for heating foods [2] and applied to materials processing [3], as well. On the other hand, application of microwave energy to metal production and recycling field [4,5] has been conducted along with the progress in microwave technologies. To the author's knowledge, its application to iron and steel production fields was first attempted 56 years later in Japan [6].

Microwave heating is suitable for exciting carbothermic reduction reactions of the ores and oxide state wastes, because of its effectivity in heating of carbon and heavy metal oxides. Although it is not possible to replace the current mass production process of the metal production, authors have attempted laboratory-scaled researches on microwave applications for reduction reactions and other processes for metal production and recycling, from an idea that there are growing expectation to apply and utilize the specific characteristics of microwave heating for a small-scale metal production process and for treatment of the by-products occurring in the metal industries.

In our studies, multi-mode microwave applicator was utilized for the heating of large mass in the experimental study, and a single mode microwave applicator was positively utilized for the purpose of fundamental elucidation of the microwave heating processes. Separated electric (***E***-) and magnetic (***H***-) field heating in the microwave cavity by which it was possible to discuss heating mechanisms of substances and compounds involved in the microwave heating processes. Microwave heating mechanisms (P [$W{\cdot}m^{-3}$]: power absorbed by a matter) are expressed by Eq. 1, where ε", μ" and σ are imaginary parts of permittivity, magnetic permeability and DC electric conductivity, respectively. And measurement of the permittivity and magnetic permeability were performed at room and high temperature, which were input for simulation of the microwave heating processes. In this article, it is intended to review the authors' research results presented for these 20 years.

$$P = \frac{1}{2}(\omega\varepsilon''|\boldsymbol{E}|^2 + \omega\mu''|\boldsymbol{H}|^2 + \sigma|\boldsymbol{E}|^2) \quad (1)$$

2. Experimental

Microwave apparatus used in the studies are either multimode or single mode microwave applicator operated at 2.45GHz. Photograph of the former and schematic illustration of the latter are shown in Figure 1 (Fig. 1). Microwave maximum power of multimode applicator was either 800W (domestic microwave oven type) and 6kW (specific microwave furnace). And 1.5kW for the single mode applicator. Temperature measurement was mainly performed by an optical method using a sapphire light pipe. The other detailed apparatus and experimental conditions are described in the references [7]. Numerical simulation was performed using COMSOL Multiphysics ver.5.5.

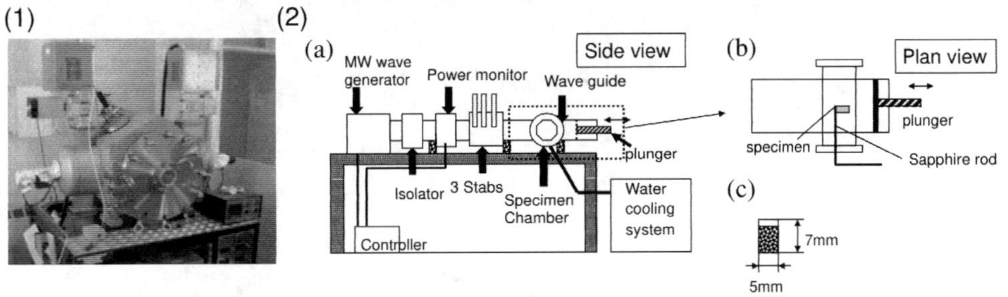

Figure 1. (1) Photograph of multimode microwave furnace (2.45Hz, 6kW), (2) Schematic illustration of apparatus [7]. (a) Side view of the apparatus, (b) plan view of the wave guide applicator and (c) scale of a specimen container.

3. Results and Discussion

3.1. Microwave processing of Ti bearing blast furnace slag, microstructural alteration and comminution

In south west China, Ti-bearing blast furnace slag [8,9] occurs because of using iron ores containing titanium which are produced in that area. It was intended to extract Ti- containing phases from the slag and utilize them as one of the TiO_2 resources by microwave treatment. If the slag is slowly cooled, slag contains Ti in a form of $CaTiO_3$ (perovskite) precipitated in the matrix. On the other hand, if it is rapidly cooled, slag is quenched into a glassy state in which all metal cations are mixed homogeneously in the glass. When microwave is irradiated (in a multimode applicator) to the glassy state slag, we observed their crystallization as shown in Fig. 2. It was illustrated that Ca and Ti are concentrated in the crystallized dendritic area by microwave rapid heating. Next, in order to separate and extract the $CaTiO_3$ phases in the slowly cooled slags, microwave cyclic heating of the slag was performed to induce some cracks. The introduced cracks can be observed in Fig. 3. Although there are many practical problems, such as the energy costs or the ability in handling large amounts, these attempts on investigation of the microwave irradiation effects were very much meaningful for our understanding microwave processing technologies in the first stage of stating our microwave research.

IOP Conf. Series: Materials Science and Engineering 1223 (2022) 012006 doi:10.1088/1757-899X/1223/1/012006

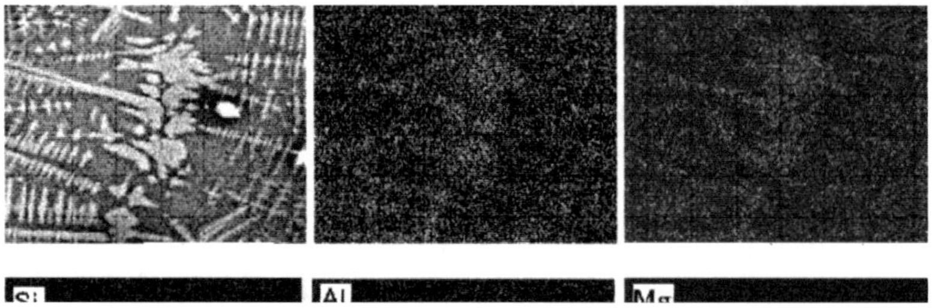

Figure 2. SEM image (left) and mapping images of Ca and Ti [8].

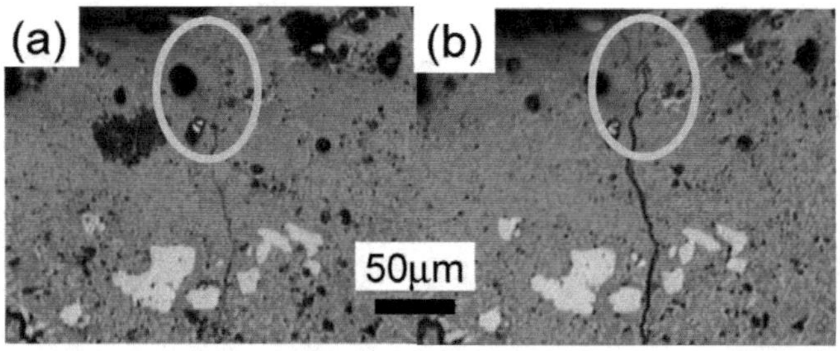

Figure 3. Optical micrographs showing extension and opening of the crack in a circled area [9].

3.2. Microwave carbo-thermic reduction of valuable metals from various oxides

Attempts on microwave heating application to the oxide reduction reaction have been performed in order to recycle valuable metals by our group. They are mostly for the steel making byproducts such as the Cr containing slag and pickling sludge of stainless steel [7,10,11]. Here, we report the reduction of NiO and Cr_2O_3 mainly, considering their microwave heating characteristics. These oxides were mixed with graphite (C), because C is not only indispensable for the reduction but also play an important role as a very good susceptor (microwave absorber). Reduction of iron oxide will also be shortly mentioned in section 4. Fig. 4 shows the microwave heating curves of NiO, Cr_2O_3 and graphite (C) in a multimode applicator [11]. Comparing with the rapid onset of the graphite heating, the two oxides exhibit an incubation time. It was discussed that this behavior is related with the large temperature dependence of the oxides' permittivity on temperature. The formation of high temperature area (hot spot) causes acceleration of temperature increase and resulted in the rapid heating after the incubation time.

In these investigations, we introduced a single mode microwave apparatus for the first places in Japan in order to obtain fundamental understanding of microwave heating of matters related with iron and steel production [7,11,12]. By this method, it is also possible to concentrate the microwave power and heat a specimen to high temperature at relatively low power input. The simulated ***E***- and ***H***- fields in a single mode microwave cavity is shown in Fig. 5. The maximum positions of the fields differ by a quarter of the wave length, such as 35mm in a wave guide at 2.45GHz. Separated ***E***- and ***H***- field heating is possible, by placing a specimen having small enough scale in each position with respect to the field distribution. The heating curves of separated ***E***- and ***H***- fields of NiO and graphite are shown in Fig. 6. Graphite can be well heated in both fields, but NiO cannot be heated well in ***H***-field. This is because NiO is not a ferri-magnetic material nor an electric conductor.

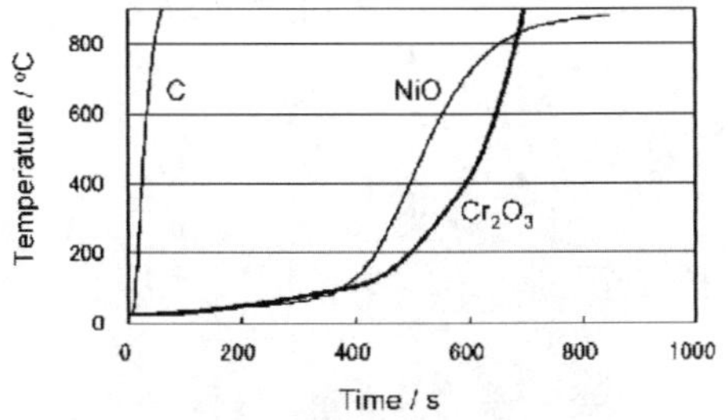

Figure 4. Microwave heating curves C, NiO and Cr_2O_3 [11].

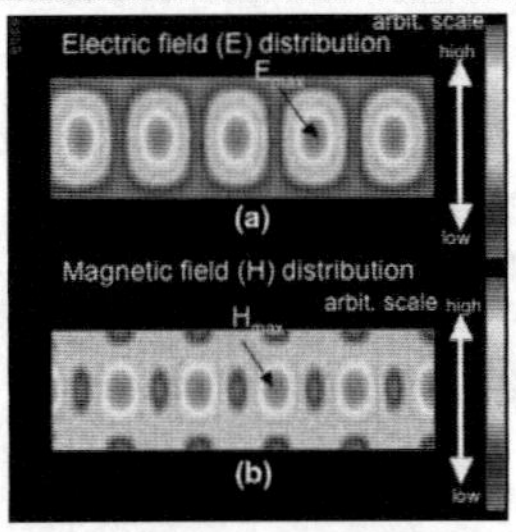

Figure 5. Simulated distributions of (a) E- and (b) H- field [7].

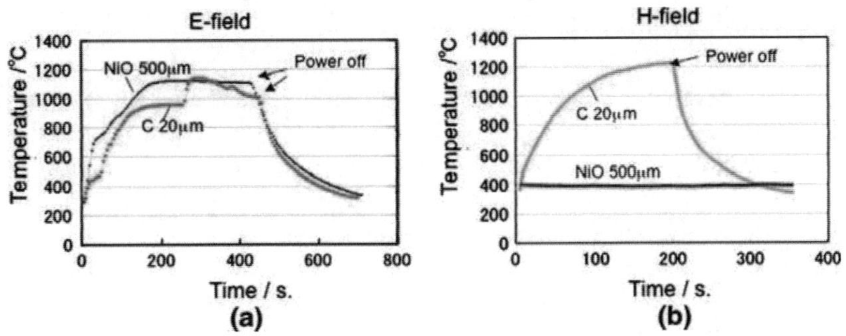

Figure 6. (1) Photograph of multimode microwave furnace (2.45Hz, 6kW), (2) Schematic illustration of apparatus [7]. (a) Side view of the apparatus, (b) plan view of the wave guide applicator and (c) scale of a specimen container.

It was observed that the reduced states of Ni are different by the reduction in ***E***- and ***H***-field. As the photographs shown in Fig. 7(a), reduced Ni in H-field maximum position consisted of particles having diameter of several micron meters. On the other hand, a dense and large Ni particle of several millimeter in diameter is formed when reduced in E-field maximum position (Fig. 7(b)). In order to account for the difference, numerical simulation on the static ***E***-field distributions around two particles are performed. Assuming the two particles either of ceramics or metals, it was predicted that the degree of the field intensification in the clearance between the particles is different, such that intensified ***E***-field is much larger between metal particles than that of ceramics, as demonstrated in Fig. 8. This tendency indicates larger possibility of discharge occurrence upon generation of reduced metals by carbo-thermic reduction. This feature is further related with the better heating tendency of NiO in ***E***-field, as well. For a case of graphite and NiO mixture in ***E***-field irradiation, first, graphite is preferentially heated at room temperature, then NiO starts absorbing microwave energy as an increase of the ambient temperature, which resulted in the incubation time. If there is some temperature distribution in microscopic scale, it is possible that the hot spot formation or preferential heating occurs, and in this area, metal generation takes place selectively, then occurrence of discharge causes generation of high temperature and melt the metals. On the contrary, in ***H***-field, graphite heating only raises the whole temperature. Although measured averaged temperature by an optical pyrometer are the same, it was anticipated the temperature difference have existed in microscopic scales.

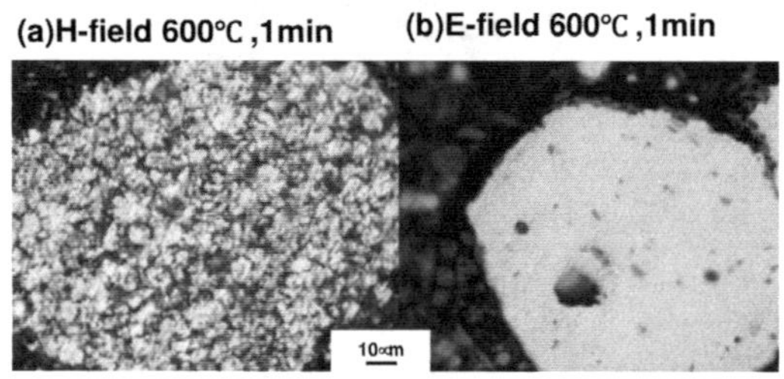

Figure 7. Photographs of (a) reduced Ni in H-field and (b) in E-field at 600°C by heating for 1 min.

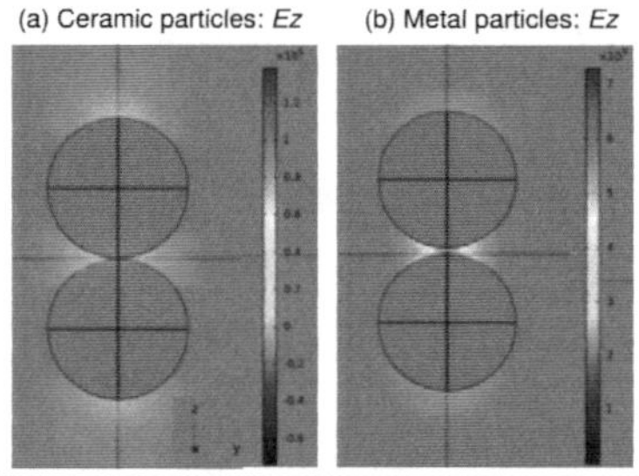

Figure 8. E-field distributions around (a) ceramic and (b) metal two particles by simulation.

3.3. Microwave drying and dehydration of wet wastes

Microwave is known to have characteristics of internal heating. In drying processes, internal heating is favorable because the temperature inside the objects could become higher than in the surface area, hence the same directions of heat and mass (vapor) transfer are accomplished. Therefore, it was intended to apply microwave heating for the purpose of drying and dehydrating of wet wastes, such as a pickling sludge [10]. We know in advance that the sludge itself can be heated up to 200°C by microwave, but dehydration is not sufficiently accomplished and the moisture still remains. Therefore, we mixed a graphite powder with the sludge and improved the microwave heating characteristics.
However, it was found that the too much addition of graphite even reduced the microwave penetration depth, as shown in Fig. 9. Color-changed (dehydrated) regions of the compacted mixture (FeO(OH), graphite (C)) by microwave heating (6kW multimode) are different with the graphite composition [10]. Namely, the larger graphite content (1:2) resulted in heating and dehydration only in the surface area. This behavior was investigated by measurement of the effective permittivity and electric conductivity of the mixture. The electric conductivity of the mixture is plotted as a function of graphite volume fraction in Fig. 10, where rapid increase in the conductivity was observed above 10%. This behavior is known as a phenomenon so called conductivity percolation.

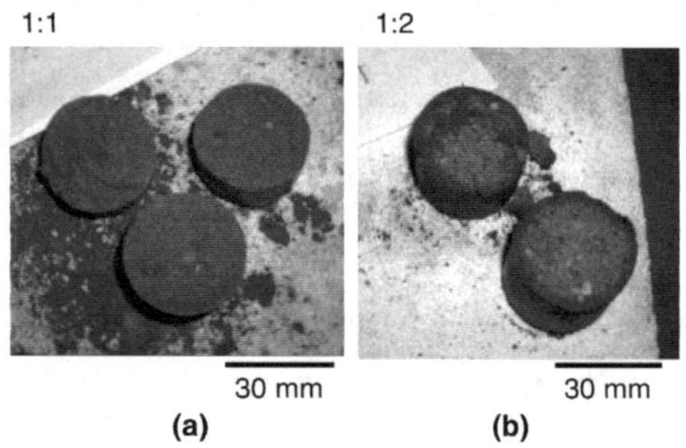

Figure 9. Appearances of the specimens of FeO(OH)/C heated by microwave. Molar ratios (FeO(OH) : C) (a) 1:1 and (b) 1:2 [13].

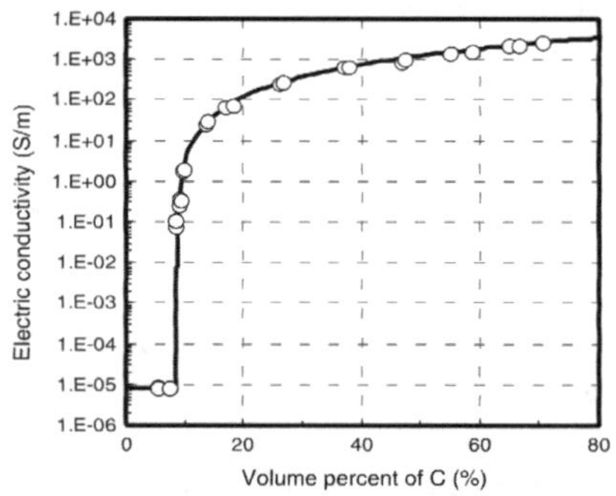

Figure 10. Effect of C fraction on electric conductivity of FeO(OH)/C mixture [13]

3.4. Vapor de-phosphorization by Microwave Rapid heating

Phosphorous is an element indispensable for fertilizers or for some chemical products. Because of running short of its natural resources, recycling of steel making slag has been taken into consideration. However, in a process of carbo thermic reduction of a phosphorous containing slag, it has to be taken into consideration that if the slag also contains iron oxides, simultaneous reduction of iron oxide occurs, from the thermodynamic analysis. Especially, when iron is reduced in a liquid state, much phosphorous is expected to be absorbed. The phosphorous cannot be separated from iron.

In order to overcome this problem, we took advantage of microwave rapid heating characteristics. It was attempted that the reduced phosphorous is to be removed in a vapor phase immediately by microwave rapid reduction, before the progress in reduction of iron oxide, which causes lowering the iron melting point by an absorption of carbon or phosphorous, leading to further loss of phosphorous. Fig. 11 indicates dependence of the phosphorous removal rate (RR) from TCP (Tri calcium Phosphate) on the heating rate (HR) (a) without and (b) with presence of iron oxide (Fe_3O_4) [14]. Removal rate is defined as the vaporized amount of phosphorous from TCP divided by the time spent to reach the target temperature. Although the removal rate is lower by one order in the case with presence of iron oxide, same tendency is shown that the removal ratio also increased with the heating.

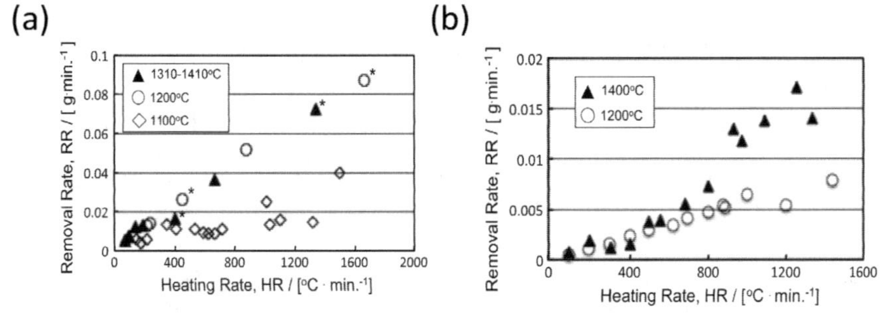

Figure 11. Relationship between heating rate and phosphorous (P_2O_5) removal rates (RR) from TCP with graphite (a) without presence of Fe_3O_4 and (b) with presence of Fe_3O_4 [14].

4. Conclusion

In this paper, our attempts of application of microwave heating to various purposes in metal production and recycling processes are reviewed. Here, we payed attention to key materials which absorbs microwave and can be heated well. Examples are perovskite ($CaTiO_3$), carbon (graphite) and the iron oxide (Fe_3O_4).On the other hand, NiO exhibited a different feature.

Carbon is not only a good microwave absorber, but also a reduction regent and very important for metal recycling. It enables rapid heating of the powder mixture, however it was suggested that optimum amount of carbon has to be selected for the microwave treatment by considering the penetration depth, as it varies with the content very much.

The studies introduced in this paper emphasized the fundamental aspects in microwave processing, however, the shown knowledge and the information are expected to be very important for the practical or industrial applications, because at present we are facing to the urgent needs to reduce CO_2 emission, and electric heating using renewable energy is one of the strategies to cope with this problem. .

Acknowledgement

The author is grateful for the financial support by Grant-in-Aid of Priority Area on Science and Technology (18070002) by MEXT Japan, and recent fundamental work was performed by Grant in Aid for general research S (17H06156), They are greatly acknowledged.

References

[1] Osepchuk J M The history of microwave heating applications, *IEEE Trans.* Microwave Theory and Techniques, **32** [9] (1984) 1200-1222.30-63.

[2] A.C.Metaxus and R.JMeredith, "Industrial Microwave Heating", IET Power and Energy Series 4, 1st Ed., The Institute of Engineering and Technology, London, UK, (1983)

[3] "Microwave Processing of Materials", Ed. by Nat. Res. Council, national Acadmy Press, Washinton D.C. USA,(1994)

[4] Pickles C A Microwaves in extractive metallurgy: Part 2 A review of applications, *Miner. Eng.*, **22** (2009) 1112-1118.

[5] Peng Z and J.Hwang J Microwave assisted metallurgy, *Int. Mater. Rev.*, **60** (2015) 30-63.

[6] Morita K, Guo M, Miyazaki Y and Sano N The heating characteristics of $CaO–SiO_2–Fe_tO$ system slags under microwave irradiation, *ISIJ Int.*, *41*, 2001, 716-721.

[7] Yoshikawa N, Ishizuka E, Mashiko K, Taniguchi S, Carbon reduction kinetics of NiO by microwave heating of the separated electric and magnetic fields, *Metall. Mater. Trans.*, 38B, (2007), 863-868.

[8] Chen Y, Yoshikawa N and Taniguchi S, Microwave heating behavior of blast furnace slag bearing high titanium. *ISIJ Int.*, **45**(9), (2005),1232-1237.

[9] Chen Y, Ishizuka, E Yoshikawa N and Taniguchi S, Crack generation in blast furnace slag bearing high titanium by microwave cyclic heating. *ISIJ Int.*, **47**(2), (2007), 193-196.

[10] Iwasaki K, Mashiko K, Saito Y, Yoshikawa N, Todoroki H and Taniguchi S, Application of microwave technique for dehydration of sludge generated in a stainless steel plant. *ISIJ Int.*, **49**(4) (2009), 596–601.

[11] Yoshikawa N, Mashiko K, Sasaki Y, Taniguchi S and Todoroki H, Microwave Carbo-Thermal Reduction for Recycling of Cr from Cr-containing Steel Making Wastes, *ISIJ Int.*, **48**(5) (2008), 690–695.

[12] Yoshikawa N, Ishizuka E, Mashiko K, Taniguchi S, Difference in carbo-thermal reduction reaction kinetics of NiO in microwave E- and H- fields, *Mater. Lett.*, **61** (2007) 2096–2099.

[13] Kawahira K, Saito Y, Yoshikawa N, Todoroki H and Taniguchi S, Penetration depth of microwave into the Mixture of Goethite with Graphite estimated by permittivity and conductivity, *Metall. Mater. Trans.*, **45B**, (2014), 212-220.

[14] Yoshikawa Y, Sunako M, Kawahira K, Suzuki K, Miyamoto K and Taniguchi K, Carbo-thermic reduction kinetics of phosphorous vaporization from tri-calcium phosphate (TCP) under microwave rapid heating with/without the presence of Fe_3O_4, Metall. *Mater. Trans.*, **49B**, (2018), 969-975.

AUTHOR INDEX